Animal Reactions to UFOs

Linda Zimmermann

Eagle Press

To contact the author, email: lindazim@optonline.net

Or write to:

Linda Zimmermann
P.O. Box 192
Blooming Grove, NY 10914

Website: www.gotozim.com

Podcast: *UFO Headquarters* on HudsonRiverRadio.com

Facebook: Animal Reactions to UFOs
        Hudson Valley UFOs
        Linda Zimmermann author page

Other UFO books by Linda Zimmermann:

In the Night Sky
Hudson Valley UFOs
More Hudson Valley UFOs

Go to Amazon.com to see the author's complete list of books on a wide variety of topics in both print and e-book formats.

Cover Photo of Cleo the rescue cat, by Raina Miller

Eagle Press, New York
ISBN: 978-1-937174-03-3

CONTENTS

Part I

Part II
Reassessing Famous and Historic Cases

Part III

Acknowledgements

Many thanks to everyone who helped with information and support, especially:

James Brown
C. Burns
Preston Dennett
Chris DePerno
Stan Gordon
Kathleen Marden
Robert Powell
Martin Willis

Special thanks to Michael Schratt for his amazing illustrations.

Thanks always to Michael Worden, my *UFO Headquarters* podcast cohost (and partner in endless adventures), and Brian Horowitz and Hudson River Radio for first giving these ideas a voice.

Heartfelt thanks to my amazing husband, Bob Strong, who helped obtain research material, edited, and put up with me during the long days of my intense "writer mode" work habits. He is a constant source of encouragement and support.

And as always, my deepest gratitude to all of the witnesses who had the courage to share their stories, because without you, none of this is possible.

Introduction

After years of interviewing witnesses, certain patterns begin to emerge, such as:

1. The more intense the experience, the greater the chance of having another.

2. Multi-generational experiences are not uncommon.

3. Dogs barking uncharacteristically are often the first sign a witness has right before a sighting.

It is the latter that propelled me down this path of animal reactions to UFOs. Clearly, they sense things that we don't, which presented a tantalizing prospect for expanding the field of ufology research. What have we been overlooking by ignoring or minimizing animal reactions?

Of course, I am certainly not the first to recognize this as at least an interesting aspect of the field of study. Most notably, Gordon Creighton created an animal reactions catalog for *Flying Saucer Review* in the 1970s, NICAP's database Category 4 is Animal Effects Cases, and Joan Woodward wrote an excellent research paper on animal reactions for the MUFON Symposium Proceedings in 2005. These were all great resources, but it seemed as though there needed to be a deeper dive into the how, what, when, where, and why these animals react the way they do.

While this is not an all-encompassing, encyclopedic reference of the subject, it is a wide-ranging overview that will, perhaps, spark discussion and further research.

At the very least, my hope is that we can seriously begin to view the UFO phenomenon through different eyes—the eyes of animals which can tell us so much if we would only stop and pay more attention.

Linda Zimmermann
November 2020

Glossary

APRO: Aerial Phenomena Research Organization was started in January 1952 by Jim and Coral Lorenzen. Currently inactive.

BUFORA: British UFO Research Association, founded in 1962. Currently active.

CUFOS: Center for UFO Studies, begun by Dr. J. Allen Hynek; continues to actively investigate and maintain an archive and library.

MUFON: The Mutual UFO Network. Currently active.

NICAP: The National Investigations Committee on Aerial Phenomena from the 1950s to the 1980s. Currently inactive.

NUFORC: The National UFO Reporting Center began in 1974. Currently active.

Project Blue Book: This U.S. Air Force program studied UFO reports from 1952 to January 1970. Currently inactive

PART I

The Case of the Paralyzed Rabbits

The tollbooth in Muscatine, Iowa

It was a very lonely job.

Al Wagner, a 34-year-old toll taker on the Norbert F. Beckley Bridge across the Mississippi River in Muscatine, Iowa, spent long hours on the nightshift in his little tollbooth. To help pass the time, he would bring carrots to feed the local wild rabbits. Even though he did this every night, the skittish rabbits would run off if he tried to get too close. However, in the early morning hours of July 22, 1981, at 3:10am, something was clearly very wrong. The rabbits didn't move as he approached.

Fearing they might all be dead, Wagner went right over to the six rabbits that were stretched out on the concrete, their front and back legs sticking out, completely unresponsive and not moving a muscle, as if they were paralyzed. Trying to comprehend what was happening, he suddenly noticed a light on the Illinois side of the bridge.

It was an egg-shaped craft with an orange glow, about 30 feet wide and 25 feet tall. As it moved toward the bridge—and toward Wagner—a yellow light turned on inside the craft. At first, at an altitude of about 350 feet, it moved in an odd step-like pattern of forward, then up, forward, then up, until it was high enough to clear the top of the bridge by no more than about 10 feet.

Norbert F. Beckley Bridge across the Mississippi River

Descending toward the water, the strange craft then veered off to the west and disappeared. As it began to move away, Wagner noticed that the yellow light went off. He also noticed what he described as a "whizzing" sound. The entire sighting lasted several minutes.

"I don't even believe in UFOs," he later told a reporter from the *Quad City Times*, "but I saw something that night that I've never seen before."

Al Wagner had also never seen the rabbits act that way before. After the craft moved off, the rabbits appeared to suddenly revive, and they jumped up and scurried off—however, without eating their beloved carrots.

These rabbits' actions are extremely important, and have serious implications for the field of ufology. However, before we discuss this further, let's look at the remainder of this case.

There were other sightings both the day before and the day after Wagner's. At 12:07am on July 21, Randy Reynolds and another man were driving near the river. They saw a "large orange disc that changed shapes." As they watched this disc-shaped craft move across the sky, it transformed into a cone shape. A third witness that night also saw the same thing.

2

UFO in Muscatine?

Rabbits' state puzzles toll-taker, center

Quad City Times, August 7, 1981

At approximately 1:00am on July 23, another witness saw a round, orange craft that "was tapered in the back" on the east side of town, which, again, was by the river.

Also, on the night of Wagner's sighting, a police officer, about a block from the police station, heard the same strange "whizzing" sound that Wagner reported. The station was only a few blocks from the river, and less than a mile from the tollbooth.

Then there was the tantalizing report of a power outage in Muscatine at 2:04am on July 22—just an hour before Wagner's sighting. Could this be both an animal reaction case, and an electrical interference case? Some reports jumped to that conclusion. However, according to Clyde Bowen of Muscatine Power and Water, the outage had nothing to do with the orange UFO, although surprisingly it did involve another animal—a raccoon who got into some of the electrical equipment that caused a short and interrupted service for about a minute. Unfortunately, the raccoon was not as lucky as the rabbits to survive his ordeal, as he was electrocuted.

In the following days and weeks, Wagner, the reluctant UFO witness, was "harassed with calls" from newspapers and television stations. The case was even mentioned on Tom Snyder's popular television show *Tomorrow Coast to Coast*, which had Dr. J. Allen Hynek as a guest. While Hynek did not mention Wagner by name, it didn't take much detective work to discover the identity of the nightshift toll taker in Muscatine. The Case of the Paralyzed Rabbits went national, and there was no escaping the media after that.

While conventional reporting systems would classify this as a case with five eyewitnesses, and a police officer who heard something unusual, it would only mention the rabbits as an amusing side note. However, perhaps we actually need to record this as a case that had 11 eyewitnesses—five humans and six rabbits. On Snyder's show, Hynek astutely stated that, "Animals are often the first thing to give a warning that something strange is going on."

This is so true! How many hundreds, if not thousands of cases begin with the witness saying, "My dog was barking like crazy, so I looked outside to see what was happening," or the "Cows and horses were in a panic and I went to check on them," and it is only then that a craft or lights are seen by the witness, who otherwise would have been oblivious to the event.

While animals can absolutely act as "early warning systems" for the approach of unidentified phenomena, their reactions can tell us so much more. Of course, animals such as dogs and rabbits have superior high frequency hearing—well beyond a human's range of hearing—and it would be easy to dismiss these cases as simply the result of frightening or irritating sounds being produced by these craft.

While it is likely that animal reactions involve more than just sound—possibly much more—even these high frequency sounds can tell us a lot. What is producing these sounds? Does it speak to some form of propulsion or some energy fields being generated?

Also—and this is very important—animal reactions prove that *something* is happening, as opposed to these events simply being the result of overactive imaginations. While we can't say exactly what these animals are experiencing, we can conclude that their reactions are:

- Honest
- Unaffected by media stories about UFOs
- Not influenced by alcohol or drugs
- Not influenced by the desire for fame or money
- Not attempting to create a hoax
- Not mistaking the planet Venus, stars, etc.
- Completely REAL and GENUINE

While many things can frighten an animal, when these reactions are accompanied by the sightings of unidentified craft or phenomena, it is safe to say the two are related. Dr. Hynek summed it up during an investigation in Ithaca, NY in 1967, which we will examine in the next chapter:

"The distress of animals is a strong point."

"Animals don't hallucinate as humans do."

What do animals perceive during these events, and what can their reactions teach us? What can we learn by reevaluating well-known cases from the past which involve animal reactions? How can we improve future investigations by considering animals as crucial witnesses? Perhaps this book will begin to answer some of those questions, as well as raise a lot more questions we need to start examining.

Here, There, and Everywhere

According to the American Kennel Club, as a breed, Great Danes "must be spirited, courageous, never timid; always friendly and dependable." So how do we explain the NICAP case of Thunder, the 175-pound Great Dane who suddenly "went crazy" the night of March 31, 1983, at 8:30pm in Sandy Hook, Connecticut?

The enormous dog was standing at the sliding doors leading to the back yard, barking—thunderously, no doubt—and acting completely out of character. The owner rushed to see what was the matter, and spotted a massive V-shaped craft over the backyard. The craft hovered silently for *a full 20 minutes*, every minute of which poor Thunder exhibited great fear and distress.

Even the next day, after the craft was long gone, Thunder was still so traumatized that he absolutely refused to go outside into the yard—and there is little a mere human can do to try to persuade a frightened, 175-pound Great Dane to do anything it really doesn't want to do. So much for being "courageous, never timid!"

Then there is the case of the 2-year-old German Shepherd, Hans, in Cresskill, New Jersey in 1968. Anyone fortunate enough to have lived or worked with one of these brave, faithful dogs knows that they will do anything to protect the humans they love.

Hans' owner and a friend were taking him for a walk near a local school, when they saw a silent, "dumbbell-shaped craft" hovering in front of them over the school field. The two people were "mesmerized" by the sight, but their "trance was broken" by Hans.

The normally "fearless" and "very protective" dog suddenly began "cowering" and desperately "trying to dig a hole in the dirt on the side of Rose Street" in which to hide. Hans was also "whimpering" pitifully, and showing every sign of being in a complete panic.

Any fascination the two people had over witnessing this amazing craft instantly evaporated, and they quickly concluded that if Hans was terrified by this object, then there was only one thing for them to do—Run!

This NUFORC case is typical of dog reactions, even though the craft is far from typical.

On September 7, 2019 a couple was driving home from Centralia, Washington at about 11:30pm on highway 12 west, toward Rochester, about 10 miles to the northwest. They saw some unusual lights, "pale yellow orbs bouncing all over the sky." Thinking they were simply the result of reflections in the clouds from a spotlight, they initially weren't concerned.

However, the lights grew brighter, and the wife realized they were on some sort of craft, which happened to be traveling the same course they were.

"Our road is curvy," she said, "and as I made the second curve, I almost wrecked the car! ...you could see 6 white lights rotating in a pattern and 4 red lights that spun, then the white lights ran to the inside or center of the craft, made a star formation, then reversed direction and made a circle around the center star-like light."

Pulling into their driveway, they could see that "the craft stopped and hovered, literally, 200 feet above us and 100 feet next to us, just above our pasture," completely silently. However, their dogs definitely heard or sensed something, and it put them in a panic.

"We got the dogs out of the kennel, and took them in the house, they had been raising holy hell, and were frantic!"

Minutes passed and the wife decided to go outside.

"I walked up the road I looked to the left and looked out at our pasture, and saw at least, 25 rays of light pointing up to the sky, they were like spotlights and were bouncing all over the place! The craft was now on the ground behind the barn so I could not see it, but I knew there were no lights out there like that! I took off for the house and told my husband we needed to leave, he agreed, and we got the dogs loaded and took off down to the next cross street, we turned left to go to highway 12 and the craft went right over us heading due west, and then it just disappeared! I am still shaken up, and just a bit afraid they will come back."

Was there a high-pitched sound being made by this craft that only their dogs could hear? Or, was something else happening that made them frantic?

Just in case the reader is beginning to get the idea that American dogs are too pampered and soft, there is an intriguing and highly disturbing example from Salto, Uruguay, which occurred on February 18, 1977, at 4am.

Rancher Angel Maria Tonna, 52, had a 3,000-acre farm, and he and his 3-year-old faithful watchdog, Topo, were bringing a herd of cows to be milked. The first sign of trouble came when the generator powering the milking shed cut out. Suddenly, a strange light appeared by the east side of the barn, and the cows became frightened and highly agitated. Tonna hurried toward this mysterious light, and Topo "aggressively" raced toward it, as well. All of the dogs on the farm also began barking in alarm.

The object was a disc "like two plates facing each other," and "glowing bright orange," which hovered over the barnyard at a height of only 20 feet. Experienced owners are well aware that when a dog is in full attack mode, almost nothing can stop it. Yet, just 15 feet from the craft, Topo stopped short, sat down, threw back his head, and began howling!

Tonna said the craft wobbled, knocked down some tree branches, and then displayed "six beams of lightning-like light, three on each side." While Topo continued to howl, Tonna experienced the sensations of "electric shocks, intense heat, and paralysis," and received burns on the arm he had raised to shield his eyes.

Unfortunately, Topo, who was so much closer to the craft, fared far worse. The poor dog refused to eat anymore and became extremely lethargic. Just three days after the sighting, the rancher found Topo dead, *on the exact spot where he had stopped and howled!* A veterinarian conducted an autopsy and found that Topo had been subjected to "extreme internal heating." This heating was so severe, "the fat on the dog's back under the skin was liquefied and then moved through the pores of the undamaged skin to solidify on the outside"—something two other veterinarians declared to be impossible according to everything they knew. But then, this was something unlike anything anyone had seen before.

It should also be mentioned that other than the horrific autopsy results on Topo and the burns on the rancher's arm, the wires of the generator had been burned out, tree branches were broken, and there was a 35-foot, circular patch of burned grass over which the craft had hovered.

This case raises so many serious questions. If this craft had been producing high frequency sounds that were irritating or painful to Topo's ears, why didn't he run *away* from the craft, instead of toward it? What force caused the aggressively charging dog to stop short and start howling? Most importantly, why on earth would Topo just sit there in what must have been excruciating pain while some possible form of radiation was literally cooking his internal organs and melting his body fat? Had the helpless dog also been paralyzed by some force exerted by the craft? Were the injuries and damage the unfortunate and inadvertent result of the craft getting too close, or was it a malicious attack?

Another case that involves paralysis of humans and animals comes from Ohinepaka, New Zealand at approximately 4:30pm in August or September of 1971. The witness's own words tell the startling tale:

"I was working on a sheep station up the Cricketwood Road, Ohinepaka, just south of Wairoa, Hawkes Bay in 1971. It was late afternoon, around 4:30 p.m. and I was heading back to the station buildings. There was plenty of sunshine and the sky was cloudless with no wind. I came through a gate on horseback with a team of four dogs and we moved onto the farm road high up on a hill. Suddenly, all the dogs 'stopped dead' in the middle of the road, just staring straight ahead with all the hair on their backs standing on end; all the barking and running about ceased. I was thinking this was odd when the horse came to a standstill as well, with its clipped mane standing on end—all five inches plus! I couldn't work out what was going on! They just stood there like statues—no barking or movement of any kind. There were no muscles twitching on the horse's neck and flanks. Then it hit me that I could no longer move my body either! I was sitting there on horseback, completely paralyzed, with the exception of slight eye and head movement. I couldn't move my body or limbs, and like the dogs, I couldn't make any sounds. It was absolutely terrifying.

8

"Then I saw what was happening! Straight down the valley ahead of us over a low ridge—lying about a quarter mile off, was a jet black, cone-shaped flying object with flashing lights rotating around its edges: red, blue, and white. One light would be red, then it would change to blue, and white. The bottom of the object was curved like a shallow bowl. It was at least 50 feet across in width and maybe 20 feet in height from the base to the top of the cone. I could see it as clear as a bell! It had no windows or portholes that I could see, and it resembled those tops that kids used to play with. The horse and dogs never moved a muscle the whole time I was watching it—it was as if they didn't even breathe.

"I had a panoramic view from my position. Initially, I was looking down on the object as it was moving down the valley toward us, but eventually it moved up higher over the hill-line, so I was able to observe it from different angles as it moved about in front of me. The object's flight path was as follows: it moved up a ridge line and hovered at the top of the hill. It then moved along the hill line about 1000 yards or more and stopped. It hovered there without moving or making any sound, but the coloured lights kept flashing around its edges in a sequence. It traveled down another ridge line, stopped dead in its tracks and hovered there. It then elevated some 500 feet or so straight up into the air above the hill and hovered there without moving. Suddenly it shot off up into the sky on an upward, curved flight path, heading south out over the sea. I have never seen anything move so fast, from zero to hundreds of miles an hour in seconds, without making any sound at all.

"About then, the horse, dogs and I all 'came back to life' and started to move again as if we had been released from a 'hold.' To me, it was an extremely frightening experience. I know very well that this thing—this craft—was not of this world. Nothing in this world that I know of flies or moves in this way without making a great amount of noise. This craft was controlled by someone or something.

"At first, I did not tell the station owner and workers, or anyone else about this incident for fear of ridicule. However, two weeks later, we noticed that a small Hereford bull had disappeared from the station and it was then that I told them what I had experienced. The bull had been in a paddock with a herd of Angus cows, situated at least a quarter mile from the road, over a small river and out of sight. Nobody except the farm employees would have known it was there. I spent three days riding around the station on horseback looking for it, and so did the owner. Not a hair of it was found. The bull didn't just go walkabout as he had too many

9

cows to keep him happy! I have often wondered whether the bull was taken by this craft—perhaps in the same paralyzed state that the animals and I had been in."

This is a remarkable case that bolsters the witness's testimony about his own physiological reactions (i.e., the paralysis) with the identical reactions by the horse and four dogs. Also, his fear was most likely shared by the animals, as is evidenced by their hair standing on end. What a sight it must have been to see over five inches of the horse's mane standing straight up!

The cone shape of the craft was also unusual as it is one of the less often seen shapes, which brings up a very important point. In 2005, Joan Woodward published an excellent paper in the MUFON Symposium Proceedings, entitled *Animal Reactions to UFOs: A Preliminary Investigation from the Animal's Perspective.* It is a fabulous resource, rich with data. However, while one of her findings didn't receive a lot of attention in the paper, it is a brief mention that packs enormous implications.

Woodward broke down the numerous animal reaction cases into many sub-categories and criteria, one of which was the shape of the craft involved. Her conclusion, based upon the data, was that *the shape of the craft had no bearing on the type of the reaction.*

Looking at just the several cases presented here so far, we have craft in the shapes of eggs, Vs, dumbbells, discs, and cones—all very different shapes, appearing in places all around the world, during sightings spanning decades. Yet, for all the variety of shapes and locales, these craft provoked the same responses of fear, pain, and sometimes even paralysis. This speaks to common forms of energy fields, sounds, radiation, etc., being produced. Regardless of exactly *what* provokes these responses, *this points to a common technology shared among all of these different shapes of craft.*

Let that sink in for a moment. There are so many wild theories about the origins of all of these craft, and one would think the dramatically different shapes could also mean very different technologies involved, which should provoke very different reactions in animals, yet, they don't.

Take the example of the massive, silent, V-shaped, boomerang, and triangular crafts of the 1980s. At the time, and up to the present day, it was often proposed that they were secret government projects. However, if they caused the same reactions in animals as all of the other shapes of craft, might we then presume that they most likely used similar technology to all of the shapes of craft?

To carry that idea one step further, this could mean one of two things. These Vs, boomerangs and triangles were built by whoever constructed everything from the discs in the 1950s, to the rectangles in the 2010s. Or, they could have been built by people who learned of this technology from these other craft, and from that, the reader can imagine all manner of scenarios.

This, of course, is a lot of speculation from which we cannot hope to derive definitive conclusions, but the basic premise stands: craft that cause the same reactions regardless of their shape most likely share a common technology, and perhaps origin.

To continue laying the groundwork for a deeper examination of animal reaction cases, let's explore more examples.

In Liège, Belgium, at 9pm on April 18, 1968, two people saw colorful beams of light projecting down from a craft. Equally astonishing, were "the sounds of general alarm which they could hear among *all the animals* in the district."[1] This is of particular interest as this must have been both a large number of animals, as well as a large variety of animals, considering they mention that it encompassed the entire district.

On Tuesday, July 2, 1968, a short time after 10pm, 24-year-old Fred Coulthard, Jr. and his 19-year-old brother, Wayne, spotted a "throbbing red light"[2] in Wooler, Ontario, Canada. Running inside to get binoculars, Fred was able to observe the rotating lights on the object as it descended and changed to a "bluish-purplish."

They noticed that the normally calm and gentle "horses in a field about 100 yards from their position were running circles during their sightings, apparently in a state of panic." Even more bizarre, were the reactions of the family's three cats.

They found one of the cats "on its back with all four legs straight up in the air. They could not bend or rouse it. Later, it suddenly snapped out of the 'trance,' ran off and was never seen again."

The second cat was also never seen again.

The third cat, one that was always very active and alert, met an untimely demise. It "was killed by a police car a short time later. The

---

[1] *Lumières Dans La Nuit*, January, 1969, p. 13
[2] *Flying Saucer Review*, Special Issue No. 2 Beyond Condon, June 1969, p.66

strange part is that the car started up, then backed over the cat, yet the noise of the car did not startle or alert the cat."

The Air Force Base at Trenton, approximately 7 miles away, had numerous calls about strange lights in the sky that night. The official excuse is that people saw a meteor. Of course, a single meteor would not stay in view long enough for Fred to go inside, get his binoculars, then continue watching it for some time. Also, would horses run in panic, and would a cat go catatonic? Would two of the cats leave their comfortable home and never be seen again because a meteor passed by?

In Franois, France on December 12, 1968, around 7pm, numerous residents saw a craft with a light which grew so intense one could not look directly at it. The craft moved through the village, and subsequently landed. People could not understand what they were witnessing. Monsieur Froidevaux and several other people also couldn't understand the behavior of Froidevaux's cat, who was inside the house, and should not have been bothered by the light.

The cat displayed "great agitation"[3] and was "meowing plaintively" during the sighting. "The cat's distress grew steadily worse, and by the time the sighting ended (it lasted for several minutes) the cat 'was literally screaming the place down and it was impossible to calm it.'"

Fortunately, as soon as the craft took off, the cat relaxed and acted normal again.

In Ibiuna, Brazil, in March and April of 1969, numerous eyewitnesses saw balls of orange light the size of car headlights. There would usually be a pair of them and "Left undisturbed, the lights would 'dance with each other'…for an hour or more and then vanish as suddenly as they had come."[4] If someone tried to approach the lights, they would disappear immediately.

Often, the lights would appear on the Fazenda Bonanza farm around 9pm, "and there was much comment on the observed fact that, just before their arrival, the pigs, dogs, geese, and fowls…would always begin to make a loud disturbance of a kind that gave witnesses the impression that these animals and birds were suffering pain and fear."

---

[3] *Flying Saucer Review,* Case Histories No. 2, 1970, p.7
[4] *Flying Saucer Review*, January-February, 1970, p.3

12

In December of 1969 and January of 1970, a flap occurred in Pudäsjarvi, Finland. There were reports of cone-shaped craft at ground level, a pear-shaped column of light, also at, or just above, the ground, and various bright balls of light.

At 1:40am on January 28, 1970, Eetu Särkelä saw a large ball of light above the Särkivaara Mountain. Four smaller balls emerged from the larger one, and while some faded from view, the others disappeared into a cloud. Eetu's brother, Valdo, had witnessed the same thing about three hours earlier. Two additional witnesses saw a large red ball of light that same night.

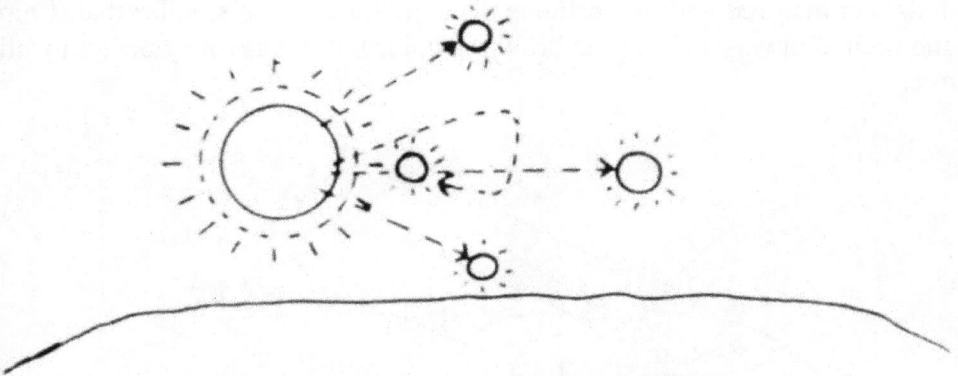

Sketch of lights above Särkivaara Mountain

At 1:30am at the Pudäsjarvi hospital, the janitor's dog "raised a loud racket and started jumping up to window-sills using its claws. Three other dog owners reported similar hysteria in their animals."[5]

On the night of January 31, more strange lights were seen, and two towns lost electrical power. Dogs in the area reportedly went "mad," including one that "was so hysterical its owner had to kill it."

The Great Ithaca, New York Flap of 1967 began the night of October 24 at about 9:30pm in the small hamlet of Newfield, just several miles to the southwest of Ithaca. While many residents just reported strange red, green, or white lights, some flashing in sequence, two boys, 12-year-old Donald Chiszar and 10-year-old Pat Crosier, claimed to actually see a craft

---

[5] *Flying Saucer Review Case Histories 2*, December 1970, p. 5-6

hovering only 130 feet above Main Street. According to these two witnesses, the craft was a silver disc, about 30 feet in diameter and approximately six to eight feet high, and tilted down at an angle in their direction.

Chiszar made a sketch of the craft and explained:

"There was a big window in the middle of the thing, which was divided into two parts by a metal bar of some kind. On each side of the bar behind the windows I could see a funny looking creature. There were two of them altogether. They stood there motionless, like robots. To the left of the funny creature on the side of the window I could also see what looked like a control panel. It was a box about 18 inches square with knobs and dials and also red and green lights. The creatures were smaller than I am and their skin was a chocolate brown in color but it was rocky or lumpy all over."

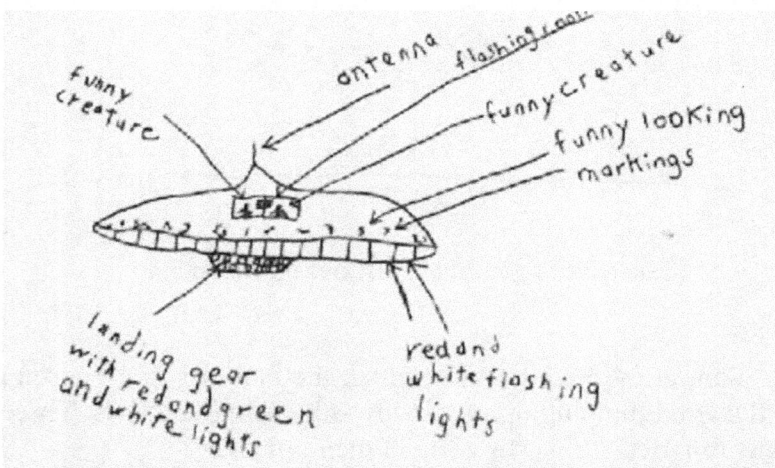

Donald Chiszar's sketch. Courtesy of the *Cornell Chronicle*.

While Crosier got scared and ran inside, Chiszar watched the object for about two minutes, at which point it "got very, very bright," "levelled off real fast and disappeared."

Just a week later, over 100 witnesses in the area had also sighted lights and glowing discs, and hundreds of additional reports continued into December. On the evening of November 9, about 40 residents gathered in the home of Mrs. Beatrice Waris to tell their stories to Air Force Lt. Gerald White, William Donovan of Aerial Investigations Research, Inc., and Project Blue Book's Dr. J. Allen Hynek.

Not surprisingly, the residents did not get any answers, and even less respect. Lt. White dismissed all of the sightings, stating "it sounds like aircraft to me." Cornell engineering students claimed the strange lights were the result of a prank. They said that they took the plastic laundry bags used for dry cleaning, rigged them with little frames containing solid fuel "Heatabs" from Boy Scout camping stoves, lit the tabs and created small hot air balloons, which upon release supposedly drifted over all the gullible residents.

Pranksters Cause Area UFOs?

By BARBARA BELL
Journal Staff Writer

NEWFIELD — The controversy about Unidentified Flying Objects, reported in the Newfield area especially, but not exclusively, for more than a month, has new fuel today.

Members of Phi Kappa Sigma fraternity at Cornell University revealed Monday that they may be the source of many "UFO's" having sent many plastic bags afloat with flaming Sterno or similar fuel during a period of about two years.

William Donovan of White Plains, president of Aerial Investigations and Research Inc., heading current UFO investigations here, said about 300 UFO sightings have been explained, mostly as conventional aircraft. Another 300 have not been explained.

He said that some of the latter may have been the Sterno plastic bags.

Futhermore, a local pilot and flight instructor, Jan Rogowicz,

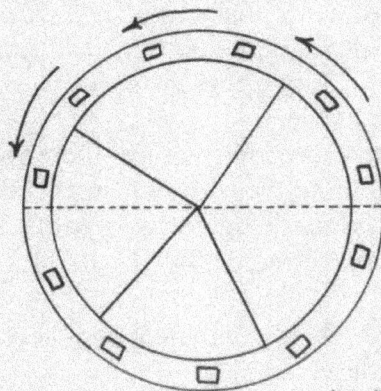

This is what Leslie Dickens of 523 N. Aurora St. says he saw hovering over his garage about 1:30 p.m. Monday.

retractable lights and lights which can be tilted in many directions," Rogowicz said. "From a distance in a night sky, this could be like an optical illusion and be confusing to someone watching below."

The object which Mr. and Mrs. Dickens report watching seemed to be 15 to 20 feet across, perfectly round, convex and silver, he said. The center portion, he said, was very bright, although there were no apparent lights. Around the rim, precisely spaced, were many windows, shaped like a television screen with rounded corners.

See accompanying diagram.

The object, he estimated, was about five foot in depth from center bottom upward.

Dickens said the hovering object appeared to be about 100 feet up. He saw no writing or lettering, no doors and there was no smoke, odor or sound.

"I believe it had to be operated by magnetic or radar propulsion," Dickens said. "It

The *Ithaca Journal*, December 5, 1967, p. 13

Dr. Hynek was initially not convinced by the reports from the witnesses, until he heard about the numerous animal reaction reports. Dogs throughout the area were barking as if in pain when the craft was near. Cats were so terrified and disoriented that "they were running into things." Cows were "moaning" and horses "crying shrilly." These were all completely uncharacteristic behaviors, which led Hynek to state:

"The distress of animals is a strong point. Animals don't hallucinate as humans do."

What was in the skies over Ithaca the fall of 1967? While we obviously can't say what it was, we can say that conventional aircraft on routine, high-altitude flights, as well as homemade, laundry bag balloons, would not induce fear and pain in the local animal population. Neither would the planet Venus, stars, swamp gas, Chinese lanterns, or weather balloons. Even if we were to discount every one of the hundreds of human eyewitness accounts, these extreme animal reactions prove that something unusual and out of the ordinary was occurring.

According to MUFON investigator Dr. Sam Greco, a retired Air Force officer, there was a *"major UFO wave which had occurred about 1800 to 1900 hours"* in the region of Williamsport, Lycoming County, Pennsylvania on the night of February 5, 1992. In the coming days and weeks, Stan Gordon, Pennsylvania state director of MUFON, said he received *"a stack of letters 8 inches thick"* describing eyewitnesses' sightings, which were of craft that were *"triangular or boomerang in shape."* Most witnesses also said the craft made a rumbling sound when it was directly overhead, which in a few instances was loud enough to rattle windows or even shake a house.

Not surprisingly with these loud sounds, there were numerous reports of animal reactions. According to an article Greco wrote for the June 1992 MUFON UFO Journal issue, "Dogs, cats, parrots and a rabbit showed signs of irrational behavior." Specifically:

- Dogs became highly agitated and began to run frantically around the house.
- The next day, two dogs refused to go outside.
- One dog was so frightened it hid under the bed all night.
- Three parrots became highly agitated and "fluffed out their feathers and perked up their ears."
- A rabbit hid in the back of its pen.

These events were so traumatizing, it took a *full week* for all the animals to return to their normal behavior.

The rumbling sound is the obvious culprit, which might make this case less important among animal reactions, even with the variety of animals involved. Also, there is the fact that one dog seemed completely unaffected by the event, and that dog happened to be deaf. However, as was stated by the human witnesses, there were no sounds heard until the craft was directly overhead, and the animals' reactions lasted longer than the rumbling sounds. Also, loud sounds alone don't necessitate panicked reactions in animals—for example, if a big truck rumbled by and shook the windows—so there still may be more going on in this case.

Then there is the case of the Romanian chickens. Of course, the word chicken itself can mean being scared, but farmers know when something unusual is happening to their flock, as in the case which occurred in Petrila, Romania in broad daylight at 2pm, on November 22,1967. The June/July 1969 issue of the MUFON UFO Journal recounted farmer Ladislau Schmit's experience:

"I suddenly saw all the chickens in my farmyard running toward me cackling like lunatics and visibly terrified. They were all flying... I raised my head and clearly saw a very brilliant object... silver or aluminum, in the shape of a disc. The object bulged slightly and the upper part was dome-like and decorated with small spikes which made me think of antennae. I called my wife... The machine was at an altitude of about 16,000 feet.

"At first...it was completely motionless in the sky, but after about a moment it began to move slowly. It soon took off at a bewildering speed toward the northwest and disappeared... Many persons to whom we pointed out the disc saw it, as well as some workers who were fixing the roof of the house in front of us."

What can "terrified" chickens "cackling like lunatics" contribute to the study of ufology? Actually, quite a lot. Now that we have touched upon the "what, where, and when" of animal reactions, keep this case in mind as we begin to explore some of the possible "how and why."

Sound

The 800-pound gorilla in the animal reactions room is sound. Some might think we need look no further than this as to the cause of the animal reactions, but let's take this one step at a time.

An average, young, healthy human's range of hearing is about 20-20,000Hz—think 20-20 for the comparisons to follow. (Of course, the older we get, that range of hearing can diminish significantly.) Dogs do not hear in the lower frequencies quite as well, but they really excel in the high range, going from about 64-45,000Hz (or even higher, depending upon the study).

It was recognized early on that a dog's upper hearing range was superior to ours, which was why Francis Galton invented the dog whistle in 1876 to facilitate training regimens. These whistles usually produce

17

sounds in the 23,000-54,000Hz range, so while dogs instantly react, their human trainers hear nothing.

As dogs are often the "early warning systems" for approaching UFOs, it would make sense for investigators to use sound sensing devices in those upper ranges. Who knows what we humans may have been missing during a sighting due to our limited hearing?

In addition to frequency, we must also consider Sound Pressure Level, which is more commonly known as decibels. Dogs are capable of hearing some sounds at fewer decibels than humans, or lower volumes, so to speak—even sounds whose frequencies are within human range. So, it must be considered that in a case where a dog alerts to a UFO, and sometime after the human witness then hears it as it gets closer, it could simply be that the dog was able to hear the lower decibel sound at a greater distance.

How many times have cat owners complained that their cats must have something wrong with them because they "just stare at nothing." Given the fact that a cat's range of hearing is an impressive 55-77,000Hz, and they are capable of hearing the ultrasonic squeak of a mouse, in a wall, in the next room, one can only imagine that cats must look at us wondering why we *aren't* staring in the direction of what they are hearing loud and clear.

Here is a chart of the high ranges of hearing for animals that are often found in reaction cases, compared to that of humans. (Note: The lower ranges vary and do not actually go down to zero. The high ends were the only numbers entered into this chart.)

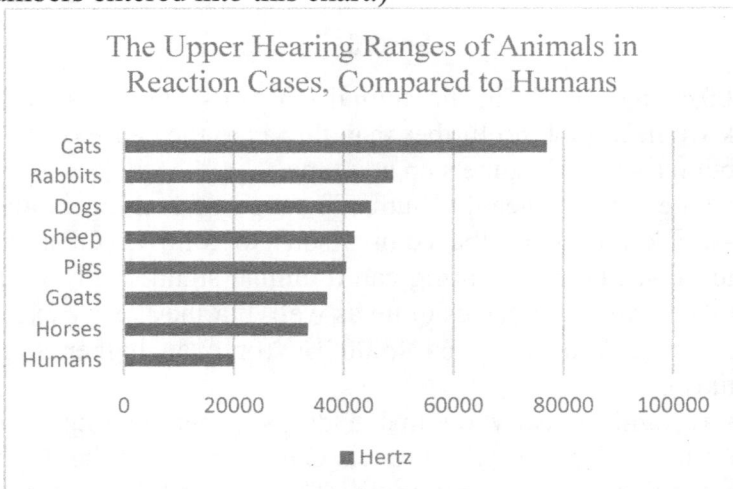

The Upper Hearing Ranges of Animals in Reaction Cases, Compared to Humans

From this, it is clear to see that the animals commonly around us hear a world that is silent to humans. It would be perfectly understandable then, if UFOs were producing high frequency sounds which irritated these animals, perhaps even to the point of generating pain. Yet, if this was solely the source of the reactions in animal cases, why did Topo in Uruguay run *toward* the craft?

In Silver Springs, Maryland on April 23, 1969, at 2am, Mrs. Virginia Guinn and a boarder awoke to the dogs "barking and howling"[6] on the farm. The boarder's dog, a German Shepherd, was "barking in a peculiar manner—a series of short barks."

Mrs. Guinn's cats were acting even stranger—in an intense, violent way. Four of them "were climbing the screen door, yowling and fighting"—behavior that Mrs. Guinn had never seen before.

The two people went outside and saw a craft "as large as two houses," that was "bluish-white…like the glow around a welder's arc." It moved past the barn in a northeasterly direction and then just blinked out as if turning off a light switch.

As soon as the craft left, the dogs stopped barking, and the cats stopped fighting and calmed down. However, it wasn't until morning that Mrs. Guinn discovered what had been going on in the horse barn. The horses had been in such a panic that night that they "had broken free of their tie-stalls...and had knocked harnesses, etc. off the walls." Also, a neighbor's horses had "torn up" their barn that same night.

Whatever happened in Silver Springs that night provoked extreme, violent, and completely uncharacteristic behavior in dogs, cats, and horses, while the two human witnesses did not report any feelings of discomfort or agitation, and heard only a brief humming sound right before the craft disappeared.

There was a curious case involving two dogs in Aveyron in southern France in 1966, where there was a farm upon which glowing balls of light had been witnessed moving near the ground. The residents called them "thinking lights," as they appeared to be under some sort of intelligent control. One night, the farmer saw one of these balls of light approaching his yard and was afraid it would try to enter the house, so he told his dogs to "Go seek 'em, go seek!"

---

[6] NICAP Animal Reactions, p. 32

The dogs immediately took off after the light and "chased it right up to the railings...in the corner of the vineyard...but never went too close to it...not nearer than maybe 1 or 1.5 meters. Then that thing disappeared in a wink, and the dogs stopped barking." Also, despite the brightness of the light, and the close proximity of the dogs, they were never illuminated by the glowing ball, so what type of light could this have been?

F. LaGarde, the author of an article on the Aveyron Case which appeared in the November/December 1966 issue of *Flying Saucer Review,* described this incident and went on to say:

"We cannot guess at the reaction of these dogs, but we have to admit solely that, at a word from their master, they chivvied the 'balls' as they would have done cattle. They did not appear scared, doubtless because they saw nothing which seemed to them abnormal, nothing which would make them hesitate to obey. This may be an important piece of evidence."

Indeed, it is most likely a very important piece of evidence, and brings up several questions. If there had been a painful high frequency sound being emitted by this ball of light, would the dogs have chased after it, and stayed as close as a meter to it? Or, were these "thinking balls" something entirely different than all the various distinct shapes of craft? Are these glowing balls or orbs pure energy, and not physical craft that create sounds or other things that frighten and agitate animals? These lights in Aveyron, at least, must be considered an exception to the belief that the majority of Unidentified Aerial Phenomena are producing high frequency sounds in a range that frighten or disturb dogs and other animals.

There is a case from Calgary, Alberta, Canada, which was described in an article written by John Magor and published in the *FSR Case Histories,* in the August 1971 issue. This case involves a horse, and took place in August of 1970 at about 8pm. Initially, the horse's reaction appeared to be from a sound that her human rider could not hear, but far more was apparently happening.

A doctor was riding his horse "along a river trail on the Sarcee Indian reserve," and the weather was clear. The doctor told investigator Bill Allan that:

"Suddenly, for no apparent reason, my horse became very alert and stiff and started to pass 'manure,' a sign of nervousness. Then very quickly it became extremely violent and practically uncontrollable, turning and twisting. I had to be very careful because the river was on my left and on my right, there were trees and a barbed wire fence."

It was only then that the doctor saw an "odd low-flying cloud ahead." He did not pay much attention to this cloud at first, because he was trying to safely get off his agitated horse, but his attention snapped to laser focus when a "solid-appearing object silently started to emerge" from the cloud.

"It was made of a material that looked like a plastic or fiberglass of blueish steel colour, or a silvery blue. The underside of it, which was all I could see, was slightly oval in shape and contained two circular vent-like structures, like the bottom of a mushroom, rotating in opposite directions. I could actually see them going round, as they were moving quite slowly."

The craft also had "a brilliant electric-blue light along its leading edge," like a "welder's arc."

"I couldn't hear anything myself, but I wondered if the object was giving off sounds that hurt my horse's ears, because by this time she was thrashing her head about very violently. I was having so much trouble with her that I was only able to catch glimpses of what was happening, but after a moment I saw the object start to go back into its cloud and the whole formation began to rise toward the southwest. It moved very slowly, not seeming to be in any hurry as it left, and the cloud became more turbulent and began to disintegrate. It trailed behind as the vehicle moved off towards the horizon and disappeared."

For two or three days afterward, the horse "was very head shy," which the doctor attributed to the horse's ears having been hurt by what he presumed had been a painful, high-pitched sound which only the horse could hear. However, a few weeks later, sores began to break out on the horse's head and neck, and she developed a "goitre-like swelling," all of which turned into what looked like "a mass of tumours." The horse's condition was captured on film by a TV crew from the Ontario Department of Education, which seems like an odd organization to have an interest in filming the growths on a horse who had encountered a UFO.

Unfortunately, the fate of the horse is unknown, as at the time the article was published, there were still tests being run to ascertain the problem, and see if anything could be done to help the poor animal.

As a side note—or is it a crucial part of the story?—after the sighting, the doctor let the horse settle down for an hour before getting back in the saddle and heading for home.

"But soon after…she started to stiffen up again and I thought, 'Oh no, not another one!' But this time she wasn't quite as uneasy and we continued along until something lying in the bush caught my eye."

Remarkably, it was another horse, "on its side, obviously dead." The horse had been "badly singed," and the air was filled with the smell of burnt hair. However, there were no signs of fire anywhere. The horse still felt warm to the touch and there was no rigor mortis, so the death had been quite recent.

The following day, the doctor and a friend went back to take a closer look at the dead horse, but it was gone. The imprint of its body clearly remained, but there were no tracks of a vehicle, which would have been necessary to remove the heavy animal.

"So unmarred was the scene, it was just as if somehow the horse had been lifted directly from the spot and hauled away by air!"

Obviously, there is a lot going on in this case, with one horse developing tumors after its terrifying encounter, and another horse found burned and dead. While high-frequency sound could have played a role in the horse's "thrashing her head about very violently," it would certainly not account for the subsequent growths—if indeed they are related, and not merely coincidence. Also, if the mystery horse's death is related, it also would not have been burned and killed by just an irritating sound. For all the questions this case raises, it does provide one answer—not all animal reaction cases are the result of sound.

A recent case documented by Stan Gordon, has both bizarre and fascinating animal reactions, and animals! It occurred on the evening of June 4, 2019 in Youngstown, Pennsylvania. While driving, two people witnessed a silent, V-shaped craft with bright white and softer colored lights about 100 yards away. Stopping to get a better look, they encountered a problem with their vehicle, as it was like "the electronic system of the car was having a seizure."

After a few minutes, they were too unnerved to stay and slowly pulled back onto the road to leave. Just 30 seconds later, a deer came out of the woods looking "dazed and confused," with another right behind it also not acting normally. Then a bird hit the windshield, but managed to keep flying. Then an animal emerged from the woods that is not even supposed to inhabit Pennsylvania.

The driver, being an "experienced outdoorsman," had no doubt that the 4-foot-long animal that stopped in the full light of the car's headlights, just seven feet in front of the car, was a wolverine! If there had been any doubt as to whether or not this was indeed a wolverine—which haven't been in Pennsylvania for generations—the case was settled when the animal bared its formidable set of teeth before rushing off.

Following the wolverine came even more deer. The witnesses were amazed and stated that "all of these animals seemed confused and frightened. They were all crossing the road as if to evade something that had scared them."

Was that "something" high-pitched sounds, the bright lights, or something else?

To add to the case of the Romanian chickens, there is the October 27, 1952 case in Gaillac, France, where Madame Daures "heard a tremendous hubbub among her chickens."[7] Thinking that there must be a hawk after them, she ran outside and "saw a huge smoke-capped cylinder and saucers."

Then there is the very odd account from Minnesota in 1976, mentioned by Loren Gross in *Some UFO Notes*. Witness Doug Fields was attempting to photograph a UFO (which is not described in Gross' excerpt), hovering over his chicken coop, with both an Instamatic and Polaroid camera. All of the photos came out black, even the ones in the Instamatic that had been taken prior to the sighting.

And while the immediate reactions of the chickens are also unfortunately not described, they did ultimately suffer unusual physical effects—four days later they had all gone bald!

In Lincolnton, North Carolina, at 6am, February 3, 1994, about an hour before sunrise, "chickens were heard to be disturbed and roosters crowed as a silent, cigar-shaped object with outward shining lights moved about the area." The 31-year-old female witness told MUFON that she estimated the craft's altitude at 200 feet, and it had strange "short tubes of light of purple, yellow, and green that shown outwards like 'stage lights.' At the rear of the object was a V-shaped purple 'corner' and two rectangular green lighted panels. The object's surface was dark and its length was estimated as greater than two car lengths. The duration of the entire sighting was indicated to be 43 minutes."

---

[7] Aimé Michel, *The Truth About Flying Saucers*, Pyramid, 1954, p. 137

The witness heard no sound. There is no mention of the chickens' behavior after the incident, but during the interview of the witness, she said she was "frightened, panic-stricken."

In Villa Rosas, Argentina, at 9:30am on July 19, 1965, witnesses saw a "blinding red light." One witness had canaries who "reacted with terrified screeches. They remained highly nervous and would not sing for two days thereafter."

In Elvia, Iowa on May 31, 1968, "baby ducks reacted" to a "car-like UFO that rotated counterclockwise." The craft shrank "smaller until it disappeared."

The cases of these chickens, the parrots in Williamsport, the bird flying into the windshield in Youngstown, and the disturbed ducks and canaries, are more than mere curiosities. Birds are key factors in animal reaction cases, because in fact, their range of hearing at higher frequencies is considerably *inferior* to humans, with chickens' upper range being a mere 2,000Hz. Therefore, if the sole source of agitation with these UFOs was intense, high frequency sound, *birds should not react at all.*

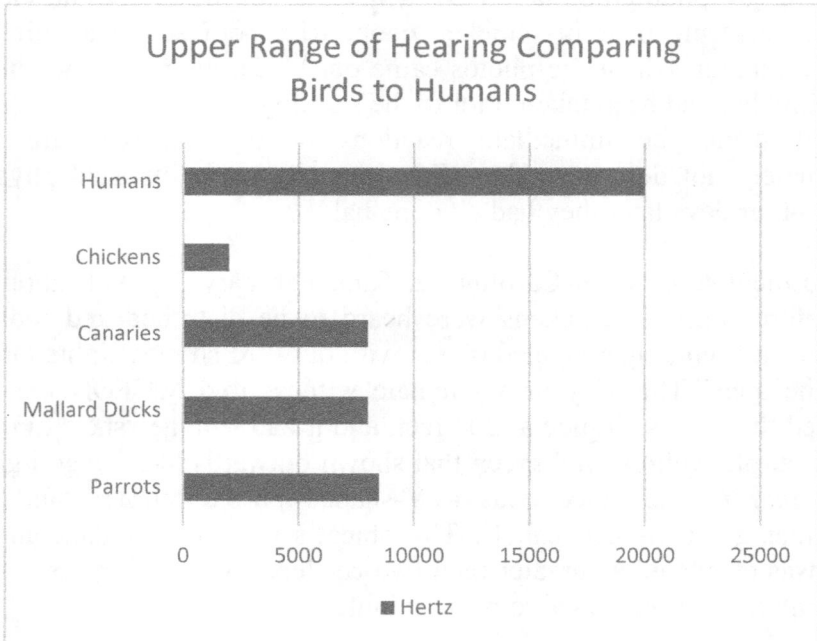

Upper Range of Hearing Comparing Birds to Humans

24

What also must be mentioned is infrasound—the range of low-frequency sound that is below humans' 20Hz. It is generally the larger animals like whales and elephants which use infrasound to communicate as far as hundreds of miles, but other animals such as rhinos, hippos, alligators, and cephalopods can also hear infrasound. Studies have determined that pigeons can hear as low as 0.5Hz!

However, when it comes to considering infrasound in the study of animal reactions, most UFO researchers do not have access to elephants and cephalopods, and little—if any—cases must exist on these animals. Of course, there is also the humble pigeon, an animal which may have a lot to teach us about the breadth of possibilities in animal reactions.

Bad Vibes?
Vibration and Pressure

Three friends were camping just 60 feet from the bank of the San Carlos Reservoir in Arizona on the clear, cold night of February 26, 1975, in an unusual case which was investigated by both APRO and NICAP. It was 2am, and two of the men were asleep inside the camper. However, a 45-year-old cement worker, Mr. G, was sleeping on the roof under the bright moonlight, as he had a respiratory problem that was exacerbated by the alcohol-burning heater inside the camper.

He suddenly awoke to one of the strangest-looking craft on record. It was "50 feet wide and 35 feet high and composed of uprights and cross pieces 4 to 6 inches in width. It seemed black in color, and had no body lights." Mr. G was able to see such detail because the object was only ten feet away from him. It was moving at about one mile-per-hour, just four feet above the ground.

As transfixed as he was by the sight, he noticed something else—the men's fiberglass fishing poles were "vibrating against the [side of the] boat" that they had beached by the water's edge. And something else bizarre was happening. "Fish were jumping out of the water 100 feet to the north and 100 feet to the south and about 50 feet from the shoreline." Mr. G pounded on the roof of the camper to get his friends outside to witness this craft, but by the time they emerged it was too late.

Mr. G.'s sketch of San Carlos Object
Sketch obtained by APRO

Mr. G managed to fall back asleep, but not for long. At 3:05am he was again awakened, but this time "by a bright light on his face and a "prickly, tingly" feeling. He put on his glasses and saw an object estimated to be 40-50 feet in diameter that was tilted at a 20-30-degree angle and had a curved upper structure." A light was coming out of a slit near the top of the craft, but then the slit closed and the object went dark. However, it passed directly overhead, and he saw "a number of holes with lighted circumferences around the outer edge."

While this was a vastly different type of craft than the one he had seen earlier, the reaction of the fish was the same—they began jumping out of the water again while the craft was near. Also, the fishing poles vibrated again, giving off a distinctive tapping sound as they repeatedly hit the side of the boat.

This is very important, as despite being two completely different objects, they both provoked the same jumping reaction in the fish and both made the fishing poles rattle. This speaks more to the similarities of the two craft, regardless of their outward appearance. But what was it the fish were experiencing that caused their unusual behavior?

An obvious clue comes from the rattling fishing poles—vibration. Fish are extremely sensitive to vibration, in part, due to the lateral line systems along their sides which they use to detect movement and vibration. Many studies have been conducted to determine whether the popularly held belief that animals can actually detect earthquakes before they strike is true. One study demonstrated that "Some species of fish possess remarkable sensitivity to pressure waves with frequencies below 50 Hz, which would enable them to sense earthquakes at least 1 to 3 Richter Magnitudes smaller than those detectable by human beings."[8]

The study concluded that "Fish also have organs which can detect slight movements of water...Thus observations of unusual behavior of fish before earthquakes may be explained if the fish are responding to small foreshocks."

And they are not the only animals to sense and utilize vibration. According to the article *Vibration and Animal Communication: A Review* in *American Zoologist*[9]:

Vibration through the substrate has likely been important to animals as a channel of communication for millions of years, but our awareness of vibration as biologically relevant information has a history of only the last 30 years. ...It has thus become clear that the use of vibration in animal communication is much more widespread than previously thought. We now know that vibration provides information used in predator-prey interactions, recruitment to food, mate choice, intrasexual competition and maternal/brood social interactions in a range of animals from insects to elephants.

It would seem as though "insects to elephants" would just about cover the full spectrum of the animal kingdom, and therefore, vibrations should be considered as potential causes of animal reactions in UFO cases—especially in birds, which are particularly sensitive to vibrations. Birds have Herbst Corpuscles, which are essentially elongated receptors. According to *Sturkie's Avian Physiology*[10] :

---

[8] Cliff Frohlich, Ruth E. Buskirk, *Can fish detect seismic waves?*, Geophysical Research Letters, Volume 7, Issue 8, August, 1980, pages 569-57
[9] Peggy S. M. Hill, *Vibration and Animal Communication: A Review*, American Zoologist, Volume 41, Issue 5, October 2001, Pages 1135–1142,
[10] Reinhold Neckar, *The Somatosensory System, Sturkie's Avian Physiology* (Fifth Edition), 2000, Academic Press

Herbst corpuscles are the most widely distributed receptors in the skin of birds. They are located in the deep dermis and they are found in the beak, in the legs, and in the feathered skin. ...In aquatic birds like ducks and geese, in some shorebirds, and in the chicken there are bill tip organs with numerous Herbst corpuscles.

Could this help explain the frantic Romanian chickens, disturbed ducks, and other birds' reactions? Are these unidentified craft producing vibrations that agitate and confuse birds, as well as other animals?

On April 28, 1961, in South Africa, Mr. A.P. Chiole's family of the farm Sushara, 12 miles from Dundee, was asleep when they heard the "most unearthly noise."[11] Mr. Chiole ran out of the house in time to see "a bright object with a fiery vapour trail pass over his house." His dogs were so terrified by the object "that one ran into a barbed-wire fence while others cowered on the ground, whining." Two hours later, a similar object, this time with lights, passed by, and even though this was at a greater distance, "the ground trembled" as it went by.

What would possess a dog to be in such a state of panic as to run into a barbed wire fence? Was it the noise, the vibrations that made "the ground tremble," or a combination of things that the Chioles couldn't sense?

During the first week of August, 1965, in Milpa Alpa, Mexico, Attilano Camacho "was awakened during the night by a great hubbub among all the domestic animals."[12] Going outside "he beheld a round, fiery object which was shooting out greenish-blue flashes of light and producing a strange vibration." Camacho and his family "watched in amazement until it finally vanished straight upwards at a staggering speed."

Other types of vibration-sensing features in animals include whiskers. Whiskers are known as vibrissae, which comes from the Latin *vibrio*, to vibrate. Unlike regular hairs, vibrissae are thicker, and their follicles are packed with sensory nerves. Cats' whiskers can detect vibrations and air currents, which help them to both find prey and escape becoming another

---

[11] *FSR*, September-October, 1961, p.28
[12] *La Nacion*, Buenos Aires, August 5, 1965, quoting *El Universal Grafico*, Mexico City (FSR files)

animal's prey. When considering the number of animals that have whiskers—from mice to manatees—vibrations must be added to the list of suspected causes for animal reactions to UFOs. At the very least, it is one more avenue to explore.

Another is pressure. According to the organization Journey North, which tracks migrations of everything from butterflies and birds to gray whales, "Studies have proven that birds are extremely sensitive to small changes in air pressure, comparable to differences of only *5 to 10 meters in altitude*." The paratympanic organ in the middle ear of birds acts as a barometer and altimeter and helps them to navigate, especially during long migrations. Also, as birds can easily sense the drop in barometric pressure before a storm, it signals them to eat as much as they can, as it is hard for birds to get food during a storm. It also causes them to seek the shelter of their nests.

Bees, fish, and many other animals are also sensitive to changes in pressure and also exhibit changes in behavior before a storm. Of course, factors like temperature changes and the ability for scents to travel farther in low pressure can also play roles in this behavioral equation, but it is clear that differences in barometric pressure alone are sufficient to elicit reactions in animals.

Even humans react to changes in pressure—think of the discomfort and pain in your inner ear and sinuses as your aircraft gains altitude. While we are not nearly as sensitive as birds and fish, as the barometer changes, people may experience things such as headaches, joint pain, or higher blood pressure.

Can these unknown craft somehow create increased or decreased air pressure? Once again, it is one of the factors that should be considered in animal reaction cases.

Where else should investigators and researchers look? The following case may provide a very important clue.

In Ramona, California, on October 15, 1974, a round craft was seen that caused numerous animals to react violently. Jeannie, Beth, Linda, and Tony Overfelt, and Pat Nelson saw the craft flying over the Santa Maria Valley, and then it appeared to land on a hillside. Slowly, the craft turned "ruby red, then intensely white." It rose up, hovered, then traveled at great speed toward them and passed overhead, with the sound being "a mix between a hum and a fog-horn."

According to Bob Gribble's report in the *Mufon Journal* #258, "No radio or TV station would work, the TV had blue spots and vertical bands. Tony's compass alternated erratically from north to northeast in one-second periods when the object was on the hillside, but when it flew overhead the compass needle lodged itself against the glass cover."

The animal reactions were even more dramatic. The NICAP report on the incident stated:

- Horses excitedly whinnied and bucked, one kicked a rail off a fence.
- One chained dog ran excitedly in and out of its doghouse, sometimes smashing into the back of the doghouse with great force.
- Another dog grasped the sleeve of a girl and tried to pull her into the house.
- Goats were reported to be "jumping around."
- Chickens were cackling and scrambling about.
- A cat ran into the side of a garage, stunning itself.

Shortly after the craft took off into the distance, the witnesses saw aircraft chasing it, which they believed to be Navy jets. Unfortunately, there was no mention of how long it took for the animal's behavior to return to normal.

This case, as well as countless other cases like it, may provide one of the most important clues in the animal reactions mystery.

Electromagnetic Fields

The Ramona case contained many intriguing aspects, not the least of which were the EM (electromagnetic) effects of television and radio interference and the compass malfunction. There are many catalogs and reports on EM Vehicle Interference Cases, where cars stall or exhibit problems when a UFO is near. These references from CUFOS, MUFON, BUFORA, the *Journal for UFO Studies*, and Keith Basterfield's report on Australian cases, represent many hundreds of examples of EM effects on vehicles from around the world. But what about the possible EM effects on animals?

Tony Overfelt's compass "alternated erratically from north to northeast in one-second periods when the object was on the hillside," clearly indicating the craft was generating a powerful magnetic field to be capable of effecting a compass at such a distance. Then as the craft passed directly overhead, "the compass needle lodged itself against the glass cover."

At the same time, a cat ran into the side of a garage with such force it "stunned itself," indicating that it was most likely both terrified and disoriented. During the Ithaca Flap, cats were running into things, also suggesting they were terrified and disoriented. In the Youngstown case, witnesses described the deer as being "dazed and confused," and a bird flew into the windshield. Many animal reaction cases similarly talk about horses and cattle running around in circles, dogs running all over the place, and other animals exhibiting signs of being impaired to the point of having no sense of direction or awareness of their surroundings. What could be the cause?

Like Overfelt's compass which exhibited the effects of EM interference, animals have their own internal compasses that they utilize every day in more ways than one can imagine. It is probably safe to declare that the following type of image has never before been used in a book about UFOs, but if a picture is worth a thousand words, then the following one speaks volumes.

According to a groundbreaking study[13] by a team at the University of Duisburg-Essen, in Germany, led by Hynek Burda, and the Czech University of Life Sciences in the Czech Republic, they "measured the

---

[13] *Dogs are sensitive to small variations of the Earth's magnetic field,*
Hynek Burda, *Frontiers in Zoology* volume 10, Article number: 80 (2013)

direction of the body axis in 70 dogs of 37 breeds during defecation (1,893 observations) and urination (5,582 observations) over a two-year period," and found that dogs universally prefer to face north. Accounting for the fact that there are subtle fluctuations in the Earth's magnetic field (MF) throughout the day, they were able to make some astonishing conclusions.

It is for the first time that (a) magnetic sensitivity was proved in dogs, (b) a measurable, predictable behavioral reaction upon natural MF fluctuations could be unambiguously proven in a mammal, and (c) high sensitivity to small changes in polarity, rather than in intensity, of MF was identified as biologically meaningful.

If one can get past the "giggle factor" of this study, these conclusions could actually be very important to the field of ufology—animals are highly sensitive to magnetic fields and can detect even "small changes in polarity." If we couple this finding with cases like Overfelt's compass reaction, then it is possible that unidentified craft are emitting magnetic fields that not only disrupt an animal's keen sense of direction, but its sense of well-being, as well. To them, such a disruption could be akin to being in an amusement park ride that spins you all around and upside down. In the presence of strong, unnatural magnetic fields, the normal framework in which animals function may become warped, distorted, and disorienting.

On November 25, 1968, Mrs. Elaine Pechy was driving in her car on Route 174 in Marcellus, New York with her 2-year-old son and their English setter dog when they saw "five, round, red blinking lights" just 100 feet in front of them.

"Then our car radio got very static," she told NICAP.

Ordinary street lights or aircraft lights should not provoke a reaction in dogs, but apparently, these were not ordinary lights as the English setter "started crying, fighting, clawing first to get out of the window, then in my lap, covering his eyes and ears, falling off the back seat..."

The EM effects continued and worsened as the car began "acting up—like it no longer had power and might be running on two cylinders instead of eight. To say the least, I was all over the road...and chugging along."

Just how out of the ordinary these lights were, became clear as the craft made a U-turn and changed from red to blinking blue and white lights, then to a white "dome-shaped object" with a "fluorescent star" next to it. The craft then somehow went inside this star-like light and they both disappeared.

Mrs. Pechy dropped her son off at her mother-in-law's home, but unfortunately kept the poor dog in the car with her as she returned to the spot of the encounter.

"I had the feeling that someone was looking over my shoulder when again the car repeated the action of losing power and the dog started to cry as if in pain again and go through the previous business, whining, barking, etc. I looked over my left shoulder. The devil couldn't have startled me more than this huge, bright fluorescent light the size of a basketball this time...If you looked at it directly it would have been blinding, like a welding torch. It also had these fuzzy lights around it."

As startled as she was, it didn't prevent Mrs. Pechy from then going home and getting a neighbor, Betsy Paranteau, to return to the site with her. Hopefully, she left her dog at home, as there is no further mention of its reactions. This time, the two witnesses watched the craft zig zagging around, but after a few minutes they had enough and they returned home with "qualmy stomachs."

Mr. and Mrs. William King of Marcellus also observed this object, as did Joan Nagan, who lived about 12 miles away in Lakeland near

Onondaga Lake—none of whom reacted as severely as the car and the traumatized dog.

At midnight in Red Hill, New Hampshire, on March 3, 1967, Mr. and Mrs. Charles Fellows and their dog were driving toward Sandwich, NH. They saw a "dark clam-shaped"[14] craft that had a motor sound and "gave off a ping-type electrical charge."

"The air had an electrical sparkling in it," and while it didn't affect the car in any way, the "dog began to be very disturbed, and his hair stood up like a brush."

The craft followed them for 20 minutes, usually no more than 100 feet above the car, but "at times, it got so close" that Mrs. Fellows "could practically reach out and touch it." Finally, the craft moved off, producing another "ping-type electrical charge," which could "be felt and heard inside the car by both occupants."

Obviously, there needs to be a correction here, and instead should have read that it was felt and heard by *all three occupants* of the car that night.

On Tuesday, November 24, 1970, at 6:45am, on his way to the Wroxham, England train station, 62-year-old Olaf Davy saw a round craft "as wide as the road"[15] hovering for three minutes low in the sky. The object had the sound of a dynamo, and Davy experienced an "uncanny, horrible feeling descending on me, like as though I was in a magnetic field…sapping the strength from me."

A friend, also heading to the train station saw the light of the craft as it moved away under the clouds. When he met up with Davy at the station, his friend asked Davy, "Hey, what the bloody hell was that?" There was also another witness of the craft in Wroxham that morning.

Davy had a previous encounter in late September of 1962 with a similar craft in the town of Kirby Bedon, which he spoke about in an interview by Peter Johnson of BUFORA.

"You'll laugh about this one," Davy said, realizing the strangeness of the story. "I'm a lover of mushrooms. I was out mushrooming in a field where there was a herd of cows…All of a sudden these cows went delirious, scampering about."

Johnson then commented that "Animals seem to sense these things."

---

[14] *UFO Investigator*, March-April, 1967, p. 6
[15] *Flying Saucer Review Case Histories 4*, April 1971, p. 1-2

Davy agreed, and continued, "…I was afraid they might trample me. I thought, 'What the hell's the matter,' then I saw a shape—no light—exactly like the one I saw at Wroxham."

"Did your hair stand on end, or anything like that?" Johnson asked.

"Well, no…but I had a pin stuck in my tunic, and I'm sure this pin was being drawn out of my clothes."

Davy went on to describe how he felt as though he was being pressed toward the ground, while this metal pin was being pulled up.

"It was a horrible sensation," Davy said.

He also described a "slight mist" and got the distinct impression "this thing was after the animals." After the craft shot straight up and away, Davy felt very weak.

Would a strong EM field have caused the sensations the witness experienced, as well as pulling on the metal pin? Would that same field have also disoriented the herd of cows, which made them "delirious" and caused them to start "scampering about" in alarm?

There was a case in Missoula, Montana in 1964 where numerous witnesses saw a "saucer landed on a hilly field." One of the witnesses reported a beam of light that shone onto his house "causing the oil furnace to start up and the ranch animals to run wild and crazy."

In the Appendix of *Strange Effects from UFOs, A NICAP Special Report*, there are several all-too-brief mentions of incidents causing both EM effects and animal reactions.

- In Black River Falls, Wisconsin on March 13, 1966, a green-white object emitting loud beeps caused interference with the radio, stereo, and television and the "dogs reacted."
- Rushville, Indiana, October 8, 1966, there was a red and white object at treetop level producing a "screaming sound." There were TV EM effects and police dogs howled.
- Alton, Virginia, November 11, 1966, a huge UFO hovered under utility lines. There was TV EM interference, a human witness was "shaken" and dogs reacted.
- Grand Marais, Manitoba, Canada, April 5, 1968, an orange object emitted a white light which caused radio interference and dog reactions.

- Lancaster, Missouri, March 10, 1969, a car drove through a beam of light from a UFO causing interference to the car and a reaction from the dog.

A more detailed account was contained in the report from New Haven, West Virginia, the night of April 17, 1967. Scores of witnesses, including Sheriff Robert Hartenbach, saw a large UFO "as big as a C-45 airplane" with two light beams moving along the Ohio River. One witness, pharmacist Lewis Summers, flashed his headlights at the craft, which appeared to respond by turning its lights on and off several times.

Three miles to the north, a boy was in a pony-drawn wagon. As the UFO approached, the horse bucked, "wrecking the wagon" and breaking the boy's glasses. The father did not believe the boy's excuse that a UFO caused so much expensive damage, until his son led him back to where the traumatized horse was still lying "with his feet sticking straight up!" Fortunately, the horse fully recovered from his ordeal.

Mrs. Lewis Capehart reported that when she saw the craft, her two German Shepherds reacted strongly, with one of them even breaking "the chain with which it was tied."

Numerous people throughout the area "reported that there was electro-magnetic interference" to their television sets, and the local radio station, WMPO, received hundreds of calls about the craft.

Could there be a better place to study animal reactions than in a zoo, where species from around the world are literally a captive audience for a curious UFO? In 1993, at the 124-acre Phoenix Zoo, at the Arizona Center for Nature Conservation, Kenneth Synnott was working as a security guard on the 4pm to midnight shift. It was a job he had held for eight years, and Ken probably expected this night would be as quiet and uneventful as all the others. Then the light went on.

About one hundred yards away, directly over the lion and tiger exhibit, was an "extremely bright" oval light—so bright it hurt his eyes. The object would also shoot down a bluish light to the ground, but that light only lasted a second. Animals throughout the zoo began making "very unusual sounds"—even those who were indoors and wouldn't even be capable of seeing the light. He immediately radioed his coworker at the front gate, but before he could even ask, the other guard said he also saw the painfully bright light.

Ken had been patrolling the zoo in the security truck, and when the light moved to a position directly above him, "the truck's motor stopped, and the trucks lights went off." Ken was naturally very frightened and "sat motionless." As the object moved away, the truck's "lights and motor came back on." Slowly, the light "drifted" toward the west and then stopped above the Phoenix Municipal Stadium about half a mile away. A few minutes later, the light disappeared, with the entire sighting lasting an impressive 15 minutes.

Ken and his partner then checked all of the exhibits to make sure the animals were not injured in any way. They appeared to be alright, and quiet, and fortunately, none were missing. Just to be on the safe side, however, they called the head animal keeper to fill him in on the bizarre occurrence. After some discussion, they all decided it was best to not say a word and avoid what, no doubt, would have been ridicule and negative publicity. The story only finally came out ten years later after Ken retired to Las Vegas.

Apparently, this UFO was noticed by more than the lions, tigers, and security guards. Jets were scrambled from Luke Air Force Base about 30 miles to the west, but by the time they reached the stadium the craft was gone.

Nothing speaks to the universality of animal reactions to UFOs than this case, where a huge variety of species from around the world all signaled that something very out of the ordinary was in their midst. The fact that the security truck engine and lights were also affected, cements the EM-animal reactions bond.

In an attempt to understand possible EM effects on animals, it's important to understand just how they sense magnetic fields. In this endeavor, birds are excellent subjects to explore, and homing pigeons are prime examples, as they have been serving as "airmail" as far back as 2900 B.C. in Egypt. The beaks of homing pigeons, and other birds, contain magnetite, which is iron oxide that is ferrimagnetic, meaning it is both attracted to a magnet, and can also be magnetized. When a homing pigeon turns its head, the magnetite particles change their alignment like a compass needle, which is obviously the perfect tool for navigation.

Magnetite is also found in the snouts of fish, as well as everything from bacteria to mammals. A recent study by Oregon State University's College of Agricultural Sciences published in the *Journal of Experimental*

Biology[16] stated that salmon use microscopic crystals of magnetite in their tissue as both a map and compass and navigate via the Earth's magnetic field." So, when birds and fish find their way back to a certain location many miles away, it is not a sixth sense, it is a magnetic sense.

In other examples, Hynek Burda, from the dog study, along with zoologist Sabine Begall, studied Google Earth images of 8,510 cattle in 308 herds from around the world, and found that cows universally prefer standing in a north-south orientation.

Caribbean spiny lobsters use Earth's magnetic field for navigation. Chickens also orient themselves to the magnetic field, and that ability becomes seriously impaired after "beak trimming," a rather barbaric practice of cutting off the ends of their beaks (which contain the magnetite particles) to prevent them from pecking one another. Whales and dolphins use the magnetic field to migrate. It has been suggested that they become "beached" when there is an alteration to the Earth's magnetic field, causing them to become disoriented.

The fox is an amazing example of using its internal compass for survival. Jaroslav Červený of the Czech University of Life Sciences in Prague, led a team of 23 researchers to study the hunting techniques of 84 wild red foxes over the course of two years. After observing "almost 600 mousing jumps," where the fox leaps into the air and pounces on its prey, they made a startling discovery. Foxes preferred to be oriented to the northeast, about 20 degrees off magnetic north, when they pounced. In this direction, they had an impressive 73% success rate. When they faced the exact opposite direction, they still caught their prey 60% of the time.

However, when the fox faced *any other direction*, it dropped to a dismal 18%! That is a dramatic difference of 55%. Imagine every time you were hungry you

---

[16] Lewis C. Naisbett-Jones, Nathan F. Putman, Michelle M. Scanlan, David L. G. Noakes, Kenneth J. Lohmann. *Magnetoreception in fishes: the effect of magnetic pulses on orientation of juvenile Pacific salmon. The Journal of Experimental Biology*, 2020

could only get a meal 18% of the time! You would quickly find a way to use all of your skills and abilities to improve those odds. This is an excellent example of just how sensitive animals are to the Earth's magnetic field, and how even slight deviations make an enormous difference.

Magnetite is not the only way animals sense magnetic fields. Back in 1978, Klaus Schulten at the Theoretical and Computational Biophysics Group at the University of Illinois at Urbana-Champaign, first predicted that cryptochromes—photoreceptive proteins found in both animals and plants—could be responsible for magnetoreception. He suggested that these cryptochromes in the eyes of birds would create a type of "magnetic filter" to overlay a pattern of the magnetic fields over their field of vision. This essentially would create road maps in the sky, as is pictured in the image on the following page.

Two studies published in 2018, one on zebra finches[17] and another studying European Robins[18], found that Cryptochrome 4, in particular, in the retinas of these birds, plays the primary role of this extraordinary ability to see magnetic fields. While birds have been the principal subjects in these types of studies, many other animals most likely share this "road map vision" of their environment thanks to cryptochromes.

Now given the role of cryptochromes, coupled with the presence of magnetite, we must again ask the question of how would animals react if a strong, artificial magnetic field were to pass overhead, disrupting the Earth's natural fields? Would they feel disoriented, off balance, perhaps even agitated and afraid that their environment was suddenly shifting in dramatic and uncomfortable ways they didn't understand? And just where does the human animal fit into this equation?

---

[17] Atticus Pinzon-Rodriguez, Staffan Bensch, and Rachel Muheim, *Expression patterns of cryptochrome genes in avian retina suggest involvement of Cry4 in light-dependent magnetoreception,* Journal of Royal Society Interface, March 28, 2018

[18] Anja Gunther, et al, *Double-Cone Localization and Seasonal Expression Pattern Suggest a Role in Magnetoreception for European Robin Cryptochrome 4,* Current Biology, Volume 28, Issue 2, P. 211-223, January 22, 2018

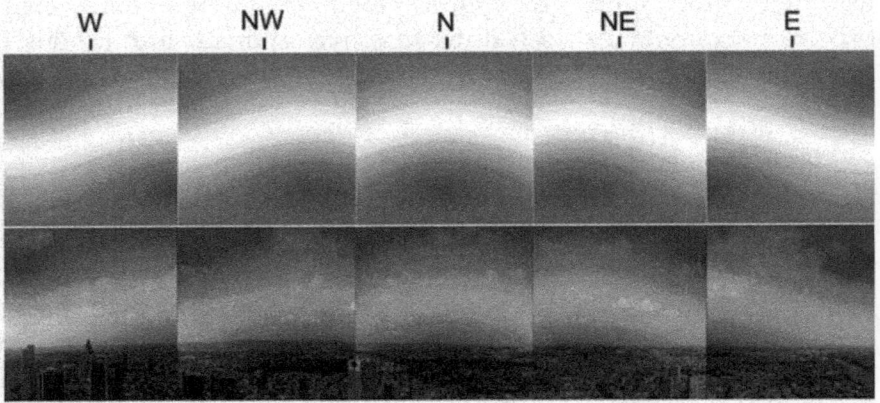

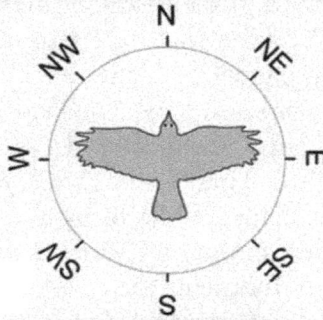

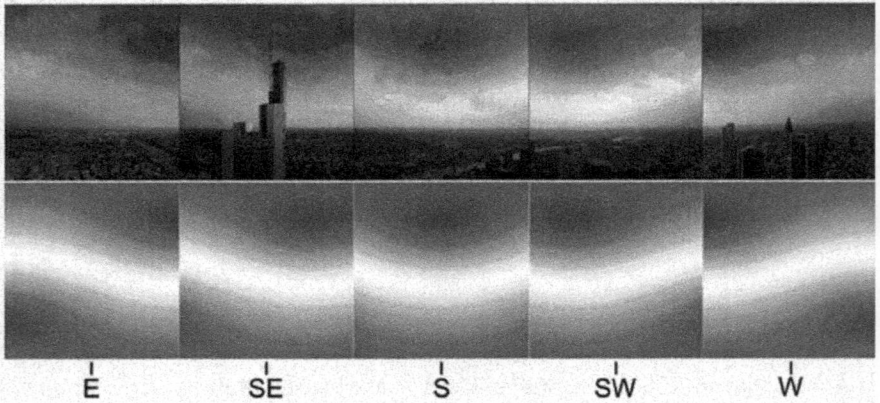

Image courtesy of the Theoretical and Computational Biophysics Group,
University of Illinois at Urbana-Champaign

Human brains contain magnetite, and human eyes have cryptochromes, the necessary "ingredients" for magnetoreception. However, even though magnetoreception is found in everything from simple bacteria to insects to complex vertebrates—which basically covers most life on Earth—the scientific community appears almost reluctant to include humans in that list, despite recent studies that show the human brain does react to shifting magnetic fields.

In a Caltech study[19] led by geoscientist Joseph Kirschvink, neuroscientist Shin Shimojo, and University of Tokyo's neuroengineer Ayu Matani, human subjects were placed in a special chamber, where all they had to do was sit comfortably for about eight minutes. The subjects' brain waves were monitored as the magnetic field was shifted clockwise and counterclockwise at intervals and durations completely unknown to the subject.

The results showed that alpha wave activity significantly increased during these magnetic field shifts, demonstrating that the brain sensed the change, even if the person was not consciously aware of those shifts. Also of note, was the fact that the subjects reactions varied considerably, with some people being far more sensitive to the change.

However, despite the Caltech study, along with other similar studies which all demonstrate some level of human sensitivity to magnetic fields, the evidence is still not there to say that modern humans have magnetoreception. Or is it?

The Kuuk Thaayore are a group of aboriginal people in the Cape York Peninsula, Queensland, Australia, and they have no words such as "left" or "right." Therefore, they will say things such as, "Please move over to the northeast so I can sit down," or "There's a stain on your southwest sleeve." Linguist Guy Deutscher studied the Thaayore language, and its speakers, and concluded that these people no doubt had "internal compasses," and were always aware of the cardinal directions.

The Kuuk Thaayore are not alone in this linguistic anomaly, either. Groups of people in Polynesia, Mexico, and Africa also utilize compass

---

[19] Wang, Connie X. and Hilburn, Isaac A. and Wu, Daw-An and Mizuhara, Yuki and Cousté, Christopher P. and Abrams, Jacob N. H. and Bernstein, Sam E. and Matani, Ayumu and Shimojo, Shinsuke and Kirschvink, Joseph L. (2019) *Transduction of the Geomagnetic Field as Evidenced from Alpha-band Activity in the Human Brain.* eNeuro, 6 (2). Art. No. e0483-18. ISSN 2373-3822. PMCID PMC6494972.

directions instead of left, right, forward, or back. Did ancient man, living in nature, and not surrounded by the EM fields created by our modern electronic age, have the same magnetoreceptive abilities all of our vertebrate cousins still possess today? And are there still groups of people around the world—as well as certain individuals—who never lost that ability, because they continued to use their magnetic sense throughout the generations?

Why this possibility is important, is because like animals who may be affected by EM fields generated by UFOs, humans may also be affected. How many times have witnesses stated that they "felt the presence" of the UFO before they saw it, or "felt off," or disoriented during an encounter? Is this the result of some mysterious sixth sense, or is it a magnetic sense? Should we not include humans as being valid experiencers of EM effects, just as we do animals and vehicles?

Something in the Air?
Odors

At 2:50am on October 4, 1969, Mr. I.A. McGregor reported that he saw "two bright orbs of light" which were following one another near Katikati, New Zealand. The sighting lasted ten minutes. Further investigation uncovered more sightings in the area over the course of several weeks, as well as evidence of possible landings. On October 8, the *New Zealand Herald* published the following:

Two mysterious circular burnt patches of grass have been found on the hillside of an open expanse of farm at Kaharoa, near Rotorua.

The owner of the farm, Mr. C.T. Johnson, of Te Waerenga Road, said yesterday he had been riding his horse when he spotted the brown coloured circles.

When he attempted to get closer his horse became "spooky and silly" and Mr. Johnson said she would not go near the circles.

The horse reared up and Mr. Johnson was forced to dismount to investigate. He said he had been riding for about an hour.

A geologist from the Department of Scientific and Industrial Research in Rotorua, Mr. D. Rishworth, accompanied a Herald reporter to the farm yesterday.

The investigation revealed one circle about 53 feet in diameter, and the other of about 30 feet. Mr. Johnson, who had been farming in Kaharoa for 18 years, stated that he had never seen anything like it before. An article in the October 10 edition of the newspaper said that a third burnt circle was also later found on Johnson's farm, and more witnesses saw UFOs.

The wave of sightings continued, and on October 16, the *Herald* reported yet another burnt circle at Puketuto, which brought the total in the area to nine at that point.

The circle, on a small island in the middle of a pond, was found by Mr. C. Blackmore on his farm. Mrs. Blackmore said her husband was herding cattle down to the pond to drink. They went so far, then turned "and belted back up the hill."

Mr. Blackmore thought the ducks had scared the cattle, but when he went down there were no ducks, just a circle on the island where the rushes had been flattened and turned brown. The remainder of the island was untouched and the rushes were quite green.

Mrs. Blackmore said the dog would not drink from the pond and further attempts to get the cattle to drink have also failed. She added that there was a funny smell to the area.

This 1969 UFO wave in New Zealand began in early September and lasted about eight weeks. Just in these two brief accounts, there are dramatic animal reactions from a horse, cattle, a dog, and ducks—who were conspicuous by their total absence. However, what is unique in these cases is that none of the reactions occurred in the presence of any UFOs.

These reactions, then, are most likely not the result of sound, vibrations or pressure, and probably not shifting magnetic fields, as the UFOs were long gone. All that remained were these burnt circles—circles which no animal would go near! Perhaps the key to this aversion for these circles comes from Mrs. Blackmore's closing comment, that "there was a funny smell to the area."

Though relatively rare, bad odors have been reported during and after sightings—usually sulphureous or chemical odors—and if humans are detecting something unpleasant or irritating in the air, what must animals, with their superior olfactory senses, be experiencing?

Another animal avoidance case comes from a farm in Finland in the winter of 1966, thanks to a letter from researcher Joel Rehnström to

Gordon Creighton. While the rest of the family slept late one night, the wife was having a cup of tea and no doubt enjoying the peace and quiet. Then a bright, elliptical light "illuminated the whole area around for several hundred meters," and it "hovered for a while above the farmer's cowshed." The light changed color several times before shooting "away at an angle."

Having previously been ridiculed for talking about a UFO the wife had seen two years earlier, she said nothing about it, until her husband came back from his morning chores in "great astonishment." It had been a particularly harsh winter and everything was covered with about 20 inches of packed snow—except the roof of the cowshed, where all of the snow had mysteriously melted! The farmer was so astonished he called the police, who came to the farm and were equally puzzled.

Apparently, however, more than melted snow was left behind, because a stray dog, who came to the house every day for food, refused to step foot on the farm! Despite his hunger, the dog "kept patrolling round the premises, but always at a distance." Was there a bad odor that prevented the stray dog from coming close and getting his meal?

In addition, the wife awoke to a "terrible headache" that morning, which was something very unusual. Also, that day, the farmer felt a "prickling sensation" on the soles of his feet, and "found strange blisters and wounds" he could not explain. Were they the result of chemicals, radiation, or something even more sinister?

It is no wonder then, if something harmful had been left behind from the UFO, that the stray dog with its keen senses chose to keep his distance.

The *Paris-Jour* newspaper described an incident that occurred in the Forest of Lond near Rouen, France on November 18, 1961. Skeptic Rémy Carbonnier, who "doesn't drink, he doesn't wear glasses and he isn't subject to hallucinations," and who thought UFOs "were a poor sort of joke," had an experience he couldn't explain.

At 2:45am, he was awakened to a green light in his room. Outside there was "something round and shining, about six metres across." It had a dome, appeared to be resting on three legs on the railroad tracks, and had shafts of orange light emanating from it. The craft took off and was gone in 20 seconds, but never made a sound.

Carbonnier was afraid to go out to investigate that night, but at dawn "he hurried to the spot where the strange thing had been." There didn't appear to be any sign of the craft to his human senses, "but Belote, his dog, sniffed the ground and ran off howling."

On September 14, 1952 in Belle Glade, Florida, a UFO circled several times over the Everglades Experiment Station[20]. An employee of the station, Gloyd Brown, reported that "an acid, ammonia-like odor" made his "eyes and nostrils smart and burn." He wasn't the only witness, however. "The cows were badly frightened by the incident, and there was a 33% reduction of milk production the following morning."

In addition to the olfactory bulb in the brain, many animals—including cattle, dogs, cats, snakes, lizards, pigs, and horses—also have a vomeronasal organ, or Jacobson's organ, located just above the roof of the mouth in a nasal cavity, which enhances their sense of smell. Humans also have a Jacobsen's organ, but there is currently controversy whether it is functional, or simply vestigial and serves no purpose.

For those who have never lived on a farm, it might be surprising to learn that cattle have an amazing sense of smell. Depending upon the wind and weather, and the nature of the scent, they can detect smells from great distances. For example, a bull can smell a female in heat from a distance of six miles! Apart from most likely being very distracting, this ability can also alert cattle to predators, sources of food, etc. And as prey animals are inherently fearful of anything unfamiliar, a strange and potent new odor at their watering hole was obviously completely unnerving.

Horses also have senses of smell superior to humans, but not quite as good as that of dogs, who have been trained to sniff out drugs, bombs, missing people, and even cancerous tumors. We can't even imagine what a dog's world must be like with so many scents swirling around them, but we can rest assured that if a dog refuses to drink from a pond, it is sensing something very unpleasant, or perhaps even dangerous.

There are many other stories about dogs, horses, cattle, and other animals avoiding possible landing sites for days, weeks, and even for the rest of their lives, which one wouldn't expect if grass had simply been flattened by swirling wind, or a small brush fire scorched a patch of ground. Like all of the other reactions discussed so far, whatever animals are sensing from these UFOs is different, unnatural, and most often frightening.

---

[20] George Fawcett, *Flying Saucers are Hostile*, Flying Saucers, February, 1961, p.15.

While the subject of this work is animal reactions to UFOs, this possible near-encounter with potential occupants described by Jacques Vallee in *A Century of Landings* is an excellent illustration of this point.

In Sherbrooke, Canada in December of 1953, Mrs. Orfei heard someone knocking loudly at her door in the middle of the night. Her German Shepherd reacted as German Shepherds always do, he lunged for the door to protect his territory. However, he quickly recoiled, began trembling in fear, and ran to hide in a corner. Needless to say, Mrs. Orfei wisely chose not to open the door.

Running upstairs, Mrs. Orfei looked out a window and saw two "indescribable" shadowy figures moving away from her house. Shortly after, about 100 meters away, a large, round craft with "blue-green lightning" rose up from the ground and took off. After reporting the incident, police found broken bushes indicating that a heavy object had indeed been on the ground.

Did the poor dog smell something horrible or threatening? Were there also high-pitched sounds that Mrs. Orfei couldn't hear? Or was it something else we wouldn't even be able to understand that terrified and intimidated a breed of dog not known for backing away from anything?

It is also interesting to note that over 100 years ago, Sherbrooke was at the center of a large wave of "mysterious airship" sightings, in 1909 along the Coaticook Valley of Quebec—sightings that have never been explained. It would be fascinating to discover if there were also animal reactions in those cases.

In any event, there is a lesson to be learned in the case of Mrs. Orfei: if your German Shepherd is terrified of whatever is on the other side of the door, don't open it.

Other Possibilities

It has long been believed that UFOs utilize some form of radiation, either long wave non-ionizing microwaves, or the higher energy, shorter wave ionizing radiation, both of which can be dangerous to humans and animals. The Cash-Landrum Case is an example of possible radiation sickness as the result of being in close proximity to an unexplained craft. On the night of December 29, 1980, Betty Cash, Vickie Landrum, and Vickie's seven-year-old grandson, Colby Landrum, were driving near Dayton, Texas when they encountered a diamond-shaped craft that was giving off a lot of heat. While there is controversy about this case—

and when is there not?—the resulting nausea, blistering skin, and hair loss that the witnesses experienced are symptoms of exposure to ionizing radiation.

In August of 1954, W.C. Hall, a sheep rancher in North Queensland, Australia, "saw six petrol tank shaped UFOs land on his ranch."[21] They emitted "oddly colored exhaust fumes" which "affected his chickens, cattle and even the jack rabbits." Hall later came to believe that there was radiation in those fumes that "brought about a change in the genes of the animal life on his ranch…as various freaks were born afterwards."

Topo the dog in Uruguay suffered "intense internal heating" to the point of his body fat liquefying, which certainly could be the result of microwaves. And did the Minnesota chickens go bald as the result of ionizing radiation? Other than feeling heat, though, can humans and animals sense radiation in some other way?

During World War II, radar technicians reported hearing an odd clicking sound, but the source could not be found. Cornell University neuroscientist Allan Frey studied the phenomena and published his findings in "Human auditory system response to Modulated electromagnetic energy."[22] What Frey found, was that people could "hear" these pulsed or modulated radio frequencies, but the response began in their brains, not their ears. The actual mechanism is still under study, but the results do show that humans can experience and sense such radiation.

In 1965, biologists at the Veterans Administration Hospital in Long Beach, California, found that cats could detect X-rays. However, they were not seeing them with their eyes; it was their olfactory bulbs that were responding! Similarly, rats were found to be able to sense X-rays, and that sense was diminished or eliminated by destroying their olfactory bulbs.

In other words, can humans and animals detect electromagnetic waves far above and below the visible spectrum? The answer is yes, although how this might relate to their reactions is yet another puzzle to be explored.

What else might cause an animal to react? Perhaps changes in temperature, or chemicals that they can't smell but can cause anxiety or

---

[21] Ibid., page 16.
[22] Frey, Allan H. *Human auditory systems response to modulated electromagnetic energy.*
Journal of Applied Physiology, 1962

alarm. Insects, birds, cattle, cats, and dogs, can all see ultraviolet lights, and some snakes see in infrared. Then there are the more mysterious possibilities, those elusive intangibles that all get lumped together by the term Sixth Sense.

The Nonreaction Reaction

On a clear night in Granville, Tennessee, at 9:30pm on April 12, 1994, "A variety of mostly farm animals were unusually calm during the time that a rectangular, rumbling object moved by at an estimated altitude of 100 feet and about 500 feet away from the witnesses. The animals' calmness persisted afterward. Animals involved were 5 goats, 4 dogs, 9 geese, 40 chickens, at least 1 cow and a horse."[23]

This case is noteworthy as all of these animals appear to have been unnaturally calm and quiet as a strange craft that was "larger than a house" moved slowly overhead at treetop level. The craft had two bright lights and a rumbling sound that was loud and clear to the two adult human witnesses. After five long minutes, the craft disappeared in a "sudden flash of white light," and the rumbling sound stopped.

If an ordinary aircraft had been rumbling over these animals at treetop level for five minutes, shining bright lights, they should have at least showed some concern and alarm. At a minimum, the dogs should have barked. This case illustrates the fact that sometimes the strangest animal reactions are those of complete nonreaction—an unnatural calm or indifference in the face of something totally foreign to their environment. (These are distinctly different from the paralysis cases where humans and animals are physically unable to move or react.)

Perhaps a clue to this calmness comes from the man and woman witnesses, who during the sighting experienced "feelings of calm and elation." There are many cases like this, where witnesses who later admit they should have been terrified by their experience, instead felt peace, tranquility, and even waves of bliss. Does whatever it is that induces these feelings in the human witnesses, also somehow sedate or manipulate the

---

[23] MUFON UFO Database, *Portion of Reports from the Files at MUFON Headquarters*, July 1999, Sequin TX, CD-ROM, Case Log #941214S by LaVere M. Pisut.

animals? And does that sedative effect linger like a drug, as in Granville where "the animals' calmness persisted afterward"?[24]

In the spring of 1967, there was a series of multiple eyewitness sightings across five states and two Canadian provinces involving UFOs that were bathing people in beams of light. On April 26, at 8:30pm, Mary Ellen Roberts was driving back from a school lecture in St. Catherines, Ontario, when she observed a "flashing red light in the sky." Her first thought was that it was a helicopter, but it was silent. She was still able to see the light when she reached her home.

"At about this point," she told a NICAP investigator, "I noticed the neighbor's dog sitting very quietly. . . looking up into the sky. We were both very still and the object came lower."[25]

The flashing red light turned to a steady green, and "Soon after this a beam of light was shot down from the object. The animal was completely engulfed in the bright glow and Miss Roberts was at its edge."

Once again, this dog should not have been "sitting very quietly," as it was something that was completely out of character.

This unusual case is something of a mixture of reaction and nonreaction. On July 19, 1965, at about 5:30pm, Denis Crowe, a former British diplomat, was walking along a beach in Vaucluse, Australia, when he saw a greenish-blue glowing, metallic, disc-shaped craft that had landed and was resting on three legs. The craft was about 9 feet high, and 20 feet in diameter. Crowe was able to get within about 50 feet, and then the craft lifted up and disappeared into the clouds.

The only other witnesses were "about a dozen or so dogs. While the object was stationary, they were all barking loudly at it,"[26] Crowe said. However, "After it took off, they were strangely silent."

Another interesting Australian case occurred at 9pm on October 31, 1967. Mr. A.R. Spargo was driving at 65 m.p.h. along a road from Kojonup to Boyup Brook in Western Australia. He was a family man and

---

[24] This single case was so dramatic, it made MUFON's Joan Woodward question whether rectangles were unlike other UFOs. For a variety of reasons, I believe they are, but that's for another book.

[25] *The U.F.O Investigator*, Vol. IV, No. 1, May-June 1967, page 3.

[26] The Sydney Daily Telegraph, July 21,1965

successful contractor, and had just finished his latest project. It was a beautiful, clear night, and he was having a relaxing drive and listening to the radio—until everything went dead and the car suddenly stopped. Mr. Spargo couldn't understand how his car went from a high speed to a dead stop without getting tossed about.

"For if you suddenly stop a car that's doing over sixty, it's like getting shot out of a slingshot, but I found I wasn't moving, yet had no sense of having stopped. All I knew was that I was the focus of a brilliant beam of light,"[27] Spargo said.

That light was coming from a 30-foot, mushroom-shaped craft "the color of bluish lightning" that was about 100 feet above him. The beam was like a "tube of light" that was at about a 45-degree angle, coming out of the bottom of the craft, and he "felt compelled to look up the tube."

"But I didn't feel any fear, and I don't remember thinking of anything in particular. I just sat and looked up that tube of light. After about five minutes it was switched off—just like someone switching of an ordinary electric light."

The color of the craft then darkened, and it took off "at a terrific speed." Spargo then found himself driving at a high rate of speed on the road again, although he had no recollection of restarting his car or starting to drive. When he got to Boyup Brook, he also found that his excellent Omega Chronometer, which never lost a minute, had somehow lost five minutes during the trip—the same period of time that his car had stopped. And the car and the watch were not the only things that stopped during the encounter.

"You hear frogs, crickets and all sorts of sounds if you stop in the bush at night," Spargo said, but during those five minutes, inexplicably, there was *complete silence*. Spargo immediately reported his experience, and later found that several other witnesses in Kojonup also saw the craft at about the same time.

While this is fascinating, the question must be asked—did all of the frogs, insects, and other animals fall silent, or was Spargo in some trancelike state where he temporarily lost the ability to hear? This was also a question Thom Reed asked himself about his family's famous encounter on September 1, 1969, in Sheffield, Massachusetts. There were numerous witnesses that night to a craft that was the size of a football field.

---

[27] Joanna Hugill, *A Tube of Light*, Flying Saucer Review, July/August, 1968, p.15-16.

Reed has been outspoken about the encounter, which he, his brother, mother, and grandmother experienced while driving in their car. He even managed to get a monument dedicated to the event, and Governor Barker of Massachusetts officially recognized it as an "off-world incident"! There is so much to this case which is beyond the scope of this book, but he discussed the part relevant to this chapter on Martin Willis' *Podcast UFO* on September 22, 2020.

A bright beam of light from a craft lit up their car, and everything around them went silent. As soon as that light was turned off, there was an "eruption of katydids and crickets and frogs. Everything came back to normal, you know, the normal sounds of the night."

Thom wisely asked himself the question as to whether the sounds of the night had genuinely been "muted," or were they simply unable to hear anything while the beam of light was on them? However, they did hear other sounds, such as "a tapping like an MRI," and sounds under their car, so he knew they hadn't lost the ability to hear. He had to conclude that "whatever field affected us" also "affected the surrounding area" and every living creature in it.

The following account is from C. Burns, who runs *The Pine Bush Anomaly Archive*[28]:

In 1995 I started my many years of trips up to the Pine Bush area in search of the UFO phenomenon, in one aspect or another. Back in the '90s it was all about putting myself in the environment that the UFO-related events were said to happen, and ground zero was West Searsville Road in Montgomery, the location said to be the best bet to experience something unusual. Back then the road was bordered by gorgeous, largely undeveloped fields on one side, and hilly forests on the other. Those who came to skywatch would usually sit close to where West Searsville met Hill Avenue, the sitting location marked by a big log.

Some Friday evenings I would have the road to myself for a little bit, at least a half-hour before being joined by at least one other person. It was just me attentively listening to the cacophony of insects and decompressing. At least once while alone, the steady drone of insects suddenly stopped, and remained quiet for a good five seconds until it

---

[28] www.pinebushanomaly.com

slowly built up again over the course of 10 seconds or so. This did not repeat in an evening, and there seemed to be no noticeable stimulus to trigger the insects to cease their activity.

I should point out that this was not a change of my perceptions, of which there are a handful of cases from West Searsville Road, that suggest witnesses went into a type of fugue state. While the insects were quiet, I could still hear distant traffic and the breeze rustling the vegetation. The second or third time this happened, I was not alone on the street, and I remember being the one who pointed out to my fellow skywatchers that things suddenly got very quiet.

I grew up in a similar area where plenty of undeveloped land played host to the amount of summer insects as in the fields off of West Searsville Road, but not once during those hours and hours outside did I ever notice a sudden ceasing of insects, and certainly not in the sharp manner they did during those evenings on West Searsville. Along with several other phenomena on that street that repeated over the few years I was able to sit and observe, those incidents remain a curiosity to me.

It is important that Mr. Burns noted that during these times when all the insects stopped making noise, he "could still hear distant traffic and the breeze rustling the vegetation." Clearly, something was going on in this very active UFO hotspot that was out of the ordinary.

A more exotic example comes from three Brazilian businessmen who were crocodile hunting very early one morning in June of 1959 in Rio Pardo. Two of them were in a boat using flashlights to look for crocodiles, when "a bluish light...approached them rapidly."[29] The light changed to green, and then orange, and as it drew within 100 meters, they could see it was a metallic saucer, about 60 meters in diameter, with "a cylindrical-shaped portion" that then slowly extended from the bottom. The craft then tilted, so that this cylinder was facing them, and they got the "distinct feeling that they were under surveillance."

At this point, "they suddenly noticed a most extraordinary thing. As everyone knows, a tropical forest is never silent. Many insects, especially cicadas, keep up an infernal din in the Brazilian jungle, even during the

---

[29] Gordon Creighton, *Mysterious Physiological Effects of Flying Saucers*, Flying Saucer Review, July/August 1967, p. 5-6.

hours of darkness. But now, for a brief time, all these noises were cut out; not a cicada or any other insect was to be heard."

When the third member of their party—who had not seen the saucer—later rejoined them, he was very upset "because of the awful silence that there had been in the forest."

How disconcerting must this be, to have the incessant "infernal din" of the jungle suddenly fall silent? And what was the scope of this effect if not a sound could be heard throughout the entire area? Crickets fall silent when someone approaches and they feel threatened, but what was the source of this apparent threat, that caused countless species to go silent?

In *Passport to Magonia*, Jacques Vallée presents a woodland silence case that may be interpreted in at least four ways.

On the rainy night of December 30, 1966, a nuclear physicist, his wife, and their two sleeping children were driving near Haynesville, Louisiana at 8:15pm, when they saw a dim red pulsating light that turned to a bright orange. The light then changed again to such an intense white that it woke up the children. The light was somewhere in the woods, "at, or close to, ground level." Very curious, but worried for his family's safety, he kept driving.

However, the next day he returned with a scintillometer, which measures small fluctuations of the refractive index of air, and determined that the light had been positioned about a mile away. As he walked toward the area, where he found burn marks, "he noticed that for some distance around the spot where the source of the light had been, animal life had simply vanished. There were no squirrels, no birds, even no insects—and as a hunter, he was quite familiar with the Louisiana fauna."

Additionally, he also found locals who had seen the light. Some farmers also claimed that they had an "important loss of cattle" during that time.

Was this a case of all the animals in the woods being frightened off by something emanating from the source of this light? If so, it is unlikely that the cattle were at liberty to take off on their own.

Were the squirrels, birds, and insects still present, but had reacted to the object that night by falling silent, and still remained so the following day?

The third option is that whatever this was had produced something that killed all the animals and insects—but, if the cattle had simply died, their bodies most likely would have been quickly found.

The fourth option does not involve animal reactions, rather it would fall under the category of animal abductions. You may be inclined to believe this fate for the cattle, but it is highly doubtful that every squirrel, bird, and insect could have been taken from a wide area of forest.

Perhaps the answer, if there is one to be found, is some combination of all four. In any event, this is another compelling case of UFO-related silence, where there should be the singing of birds, buzzing of insects, and squirrels chasing each other and running around gathering food.

Mr. and Mrs. Robert Miedtke were staying in their camping trailer while visiting some relatives near Ogema, Wisconsin, on the night of August 12, 1967. At 2:30am, their German Shepherd began barking, but his barks slowly faded and he became quiet. In fact, *everything* became quiet—no crickets, frogs, or any of the usual night sounds.

Looking out the window of the trailer, they "saw a half-moon-shaped object near or on the ground in the pasture. From the object emanated a white beam of light which extended 150 feet to the milk house which was 25 feet from the trailer."[30] For an astonishing two and a half hours, the object beamed this "fluorescent light tube," but the Miedtkes were too afraid to leave the trailer to investigate, or even to look out the window more than a couple of times for fear of being seen.

They were even more frightened when they heard heavy footsteps walking on the gravel around their trailer. This happened three separate times over the course of an hour. During the entire two and a half hours, that was the only sound they heard as otherwise there was "dead silence outside." Their dog would normally have been very protective of the Miedtkes, especially if someone or something was walking around the trailer, but he didn't make a sound, and neither did any other animal or insect that night.

At dawn, their "dog began to whimper and finally to bark," and all of the "usual night sounds" returned. They looked out the window and saw that the craft was gone. They were to find out that one and a half miles away, another farmer's dogs became "very excited" at 2:30am. The farmer was unable to see the craft from that distance, but it is possible his dogs heard or felt something—although obviously not whatever it was that silenced everything around the Miedtke's trailer that night.

---

[30] *APRO Bulletin,* September/October 1967, p. 11

The Attraction Reaction

In rare cases, animals are actually drawn to UFOs and unusual activity. They differ from the uncharacteristic passivity of the nonreaction cases as there appears to be a genuine eagerness to approach an object (in a completely nonaggressive manner), or remain in the area of a sighting.

The following is an excerpt from my book *Hudson Valley UFOs*, and involves an incident from July of 2013. I was with my husband, Bob, our friend, Art, and friend and Pine Bush UFO archivist, C. Burns. We were enjoying a beautiful night of sky watching at the home of Ginny, who had property overlooking the most active hotspot in the Pine Bush area. We set up chairs in the field and had a wonderful evening. Then:

It was getting close to midnight when Art said he would be leaving, as he had to get up early in the morning. He was standing with Bob to his left, and C. Burns was to Bob's left. I was standing facing the three of them, and Ginny was seated behind them. I was going to kid Art that as soon as he left, that would be when something would happen, but I never got to finish my words.

"You know what's going to happen as soon as you leave, don't you?" I began. "That's when—"

I stopped short because the three of them suddenly lit up brightly for an instant, as if someone had just taking a flash picture behind me. The only trouble was, no one else was with us. My back had been to a section of woods a couple of hundred feet away, and that's where Bob and C. Burns said the flash had originated—inside the woods. Normally, that would be an interesting little anomaly that we would take note of, and then mentally file it away.

However, it was a different matter here, because in Pine Bush a lot of the activity is at ground level, and has often been described as a strobe light or flash bulb effect. C. Burns has experienced this on many occasions in Pine Bush, and he had previously tried to describe the effect to me, but I just couldn't quite picture it. I could certainly picture it now!

Art also wrote a statement saying the flash reminded him of an M-80 firecracker exploding in the woods, but with absolutely no sound. I personally experienced a strange, tingly, strong adrenaline rush-type of reaction and barely slept a wink that night.

The woods on the edge of the field where the flash originated,
taken from where we were sitting that night.

Also, when I sent Ginny a thank you email the next day, I mentioned that it would be interesting if she took her dog to the field where we were sitting, and into those woods to see if he had any reaction. She replied that her cats had acted very strangely that night after we all left. They had been on the field with us, and every night they always followed Ginny back into the house—except this night.

"I couldn't get the cats to come off the field. They followed me half the way back to the house, but that was it...they turned around (!) and headed back to the field."

That had never happened before. What would keep the cats from going inside, as they always had done before? What was it about the field that made them want to stay?

In Valparaiso, Florida on October 24, 1988, at 9:30pm, a 30-year-old woman was at home watching TV when she saw a light outside. Turning out her own lights, she then went out to the terrace and saw a dull gray rectangle about 100 feet long, 3 feet high, and 4 feet deep.[31]

The craft had "3 bright lights at one end" and was moving so slowly that it often appeared to be hovering. It "took a complex flight path down, across, and back up the far side of the Boggy Bayou. Its altitude was treetop height and its distance from her at the closest approach was 100-500 feet." During the five-minute sighting, the witness' "dog and a cat

---

[31] NICAP Animal Effects Case #881024

exhibited excitement and curiosity and both wanted to go toward the object."

The witness herself stated that she felt "feelings of calm and elation, at the time" but did not experience any sort of sedative effects. Unfortunately, that elation did not persist, as "in following days" she experienced a "skin rash, some hair loss, nausea, and muscle aches." There is no mention of any subsequent behavioral or physiological effects in her dog or cat.

Summary of Part I

Given all of these examples from around the world, and considering the many possibilities of how and why animals react to UFOs, what should be becoming clear at this point is that animals are an extraordinary resource for studying UFO phenomena, and that through their reactions, we may be able to gain deeper insight into the nature of those phenomena. At the very least, animals are the unbiased, impartial witnesses that just might be able to stick a pin in the skeptics' and naysayers' theories on many famous and historic cases.

The controversial *Madonna with Saint Giovannino,* a 15th-century painting attributed to Domenico Ghirlandaio. Some see a UFO in the sky with a man and dog watching it, while art historians claim it is simply a representation of an angelic presence, or a star. Note: it is worth finding a high-resolution color image to bring out all of the details.

PART II
Reassessing Famous and Historic Cases

Animal reactions to UFOs have certainly not been ignored, but have researchers given them the attention they really deserve? Can we now look at past cases with different eyes—the eyes of animals and their superior senses? With a new and expanded understanding of these amazing and often underappreciated witnesses, what will a reassessment of these cases teach us, and will they offer evidence that might even thwart the skeptics' so-called explanations?

In Delphos, Kansas on November 2, 1971, at about 7pm, 16-year-old Ronnie Johnson was out in a shed by the sheep pen finishing up some chores before dinner. With him was his beautiful dog, Snowball; a large, white dog who helped herd the sheep. It should also be noted in regard to those sheep, that this "event took place during the time their lambs were being born."[1]

Ronnie Johnson and Snowball

---

[1] NICAP Animal Reaction Case Directory, 771102

Suddenly there was a rumbling sound, which Ronnie later described as sounding like "a washing machine out of balance." Leaving the shed to investigate, he and Snowball saw an 8-foot-wide "toad stool"-shaped craft just a few feet above the ground, and about 75 feet away. It was very brightly lit, and had red, orange, and blue lights.

A depiction of the craft at Delphos, courtesy of Michael Schratt.

Rather than bark at the strange object intruding in *his* territory, Snowball uncharacteristically just stood there quietly. His "ruff was not up" and he had no outward signs of alarm—he just calmly looked at the object. The sheep began "bellowing," but seemed unable to move. Ronnie then realized that he and Snowball were also unable to move. Given the time Ronnie's mother first called him for dinner before the object appeared, and when Ronnie finally ran back to the house to get his parents to also witness the craft, about 30 minutes of paralysis had passed.

60

Finally, the craft grew in brightness, "angled upward" and moved off with "a high-pitched jet-like sound." Ronnie was temporarily blinded, and when he regained his vision, he ran to get his parents, who came outside in time to see what looked like an intense car headlight in the sky speeding away.

They found some broken branches and tree limbs where the craft had been, but the truly startling thing was the approximately 8-foot-diameter glowing ring on the ground. Ronnie's mother, Erma, ran to the house to get a camera to take a picture of the ring, which she said was so bright she could have "read a newspaper" from its light. The broken tree limbs were also glowing. Naturally curious, Erma touched the glowing, powdery substance, and her fingers went numb, a condition that persisted to some degree for the rest of her life.

Hundreds of pages have been written about the chemical analysis of that powder and the soil, which turned out to be extremely hydrophobic—water repelling—and the fact that nothing would ever grow there again, so that will not be covered here. Delphos is undoubtedly one of the best trace cases in ufology, but it is also one of the best animal reaction cases.

Why the animal reactions here are of great importance, is because skeptics of this case try to dismiss it as a hoax perpetrated by the Johnsons, and say that the glowing ring was nothing more than fungus—something known as a Fairy Ring. (Other excuses involve "fire balloons" and chicken excrement.) However, even if this was true, how can we explain the subsequent behavior of Snowball and the sheep?

Both Ronnie and his father, Durel, noted that after the event, Snowball looked "up in the air all the time and just cried." The following day, a local reporter, Thaddia Smith, said that both Ronnie and Snowball still appeared frightened, and confirmed that the dog "wandered around the yard with his head up in the air as if still looking for the object."

Another independent witness, Undersheriff Harlan Enlow, stated that the next day they "had to force the dog back into the area [where the UFO had hovered] and he was shivering and pawing the ground. He was upset, very agitated."

Snowball had always been a fearless, outdoor dog, but every night for more than a week he desperately clawed at the front door trying to get into the house. He had been so terrified and traumatized by the UFO that he actually damaged two wooden doors tearing into them, so the Johnsons had to start leaving the porch door open, and then later installed a metal door.

61

Snowball's behavior got even stranger. As he was herding the sheep, Snowball inexplicably "ran full tilt into a fence." In the days and weeks following the sighting, he began bleeding from his nose, he wasn't eating much, and the poor dog clearly was not feeling well. They brought him to the veterinarian, who, under anesthesia, removed "a 3.5-inch bearded piece of probable vegetation" from up one of the dog's nostrils! There were other reports that the object was metallic, and another that it was a previously unknown parasite. Whatever it was, what was a 3.5-inch piece of *anything* doing in Snowball's nasal cavity? It took quite a while, but fortunately, Snowball recovered.

Then there were the sheep, who "would jump out of the pen after dark" every night for a week after the sighting. Clearly, they, too, had been severely traumatized by the UFO.

So, in light of both the immediate and lingering animal reactions, the questions that must be asked are:

- Were the Johnsons so clever as to train their sheep and dog to be part of the hoax?
- Was Snowball suddenly so terrified by a Fairy Ring of fungus that he tried to claw his way through wooden doors to get away from it every night?
- Was something left behind by this craft, something the humans could not detect, but something that caused panic and distress in Snowball and the sheep for over a week?
- Could any ordinary object or phenomena, something that occurred in that area on a somewhat regular basis, have been responsible for the extreme reactions of Snowball and the sheep?
- Do these independently witnessed animal reactions, in fact, prove that this was a real event, with a real unknown, that produced some sort of sounds, smells, EM fields, vibrations, etc., that were not normal or natural?

It should also be pointed out that there were two other possible human witnesses to this object. Elton Smith, an educator at a school in Bennington, Kansas (31 miles to the southeast), saw a bright streak of light descending in the direction of Delphos at 6:20pm, forty minutes before the sighting on the Johnson farm. Also, Lester Ernsbarker in Minneapolis, Kansas (14 miles to the southeast), saw a similar bright light from the direction of Delphos at 7:30pm, the time that the object left the farm.

For obvious reasons, TV shows, books, and articles about the Delphos, Kansas encounter focus on the glowing ring. However, remarkably, many of them barely mention Snowball and the sheep, who are also crucial witnesses in this case. While a dog and some sheep can't tell us exactly what it was that they experienced, these bizarre, unusual, and uncharacteristic animal reactions can, at the very least, deflate the fire balloon theories and pluck out the fungus excuses.

***

On September 3, 1965 at about 2am, 18-year-old Norman Muscarello was "thumbing down" Route 150 in Kensington, New Hampshire, hoping to catch a ride on the long walk home from his girlfriend's house. He was headed for home in Exeter on "a clear, beautiful night" with "plenty of stars" when something frightening appeared in the sky.

"I observed planes in the sky earlier," Muscarello said. "It's pretty easy for me to understand the difference between a plane and what I saw. I'm sure if you experienced it, there wouldn't be any question in your mind. I got just past the Dining farm—There's a little field on the right-hand side. You can kind of see the glow of Hampton Beach and the lights from the beach, which is distinguishable. What I'm trying to say is that you can see what's going on. It's pretty obvious that that's the beach area."

Coming from the north, he saw lights that were "pulsating erratically" headed toward him. The "speed must have been something terrific" as it came upon him in the blink of an eye. "There was absolutely no sound, other than the fact that I heard horses in Dining's field, raising holy hell, kicking the barn. Crickets seemed to just quit."

The object passed by, but then returned. Frightened and alone, Muscarello ran, "tripped on something and fell into the ditch, and I lay there with my head down." He saw that the house next to the Dining's "seemed to turn out like a blood red" from the lights of the object which were "still pulsating in erratic positions."

Muscarello then flagged down a car to bring him to the Exeter police station[2], where he didn't expect the reaction he was to receive upon relaying his wild story to Officer Roland "Scratch" Tolland. Instead of ridicule, Tolland said that he wasn't surprised as "I just had two reports

---

[2] Muscarello would not give the name of the man and the woman in that car, as the woman was not the man's wife!

before you walked in here, one from Raymond and another one from Hampton Beach," of the same unidentified object.

Earlier on Route 108, Officer Eugene Bertrand, Jr., had come upon a woman sitting in her car who was visibly shaken and upset because a "huge object with flashing red lights" had followed her for 12 miles from the town of Epping, and had hovered over her car. Now curious, Bertrand took Muscarello back to the scene where they began walking toward the woods. The horses were still extremely agitated, kicking violently and making sounds of alarm, and all of the dogs in the area had joined in, howling and barking.

Then they saw the object returning, which Betrand described as "this huge, dark object as big as a barn" with "flashing red lights on it" only 100 feet above the ground. As the craft moved silently towards them, approaching to within only 100 feet away, Betrand drew his service revolver and aimed it at the mysterious craft. Deciding that firing on this huge, silent, unknown object wasn't the smartest idea, he and Muscarello ran back to his patrol car.

Left to right: Norman Muscarello, David Hunt and Eugene Bertrand, and (seated) "Scratch" Toland

As the two awaited a second policer officer, David Hunt, they continued to hear that the animals were still in a state of extreme agitation. When Hunt arrived, he was also able to closely observe the strange craft, and agreed it was nothing conventional. After the craft finally moved off, the three men went back to the station to write their separate reports.

Naturally, this incident, as well as subsequent reports in the area in the following weeks, made quite a stir. Two officers from nearby Pease Air Force Base interviewed the three men and then stated that they were "unable to arrive at a probable cause of this sighting. The three observers seem to be stable, reliable persons, especially the two patrolmen." They "viewed the area of the sighting and found nothing in the area that could be the probable cause. Pease AFB had five B-47 aircraft flying in the area," but they did "not believe that they had any connection with this sighting."

Then the official backpedaling began, with the Pentagon falling back on the tried and true excuse that the witnesses saw "nothing more than stars and planets twinkling...owing to a temperature inversion"—even though weather conditions at the time could not have produced a temperature inversion. Project Blue Book then added that it was probably nothing more than aircraft from Pease involved in something called Operation Big Blast, even though Pease had already stated it couldn't have been their aircraft. In 2011, the *Skeptical Inquirer* published an article by Joe Nickell and James McGaha asserting that what the three men witnessed was undoubtedly the flashing red lights a Boeing KC-97 Stratofreighter that was refueling aircraft from Pease—even though Stratofreighters were not known to hover silently 100 feet off the ground.

KC-97L Refueling A-7D Corsair IIs
National Museum of the U.S. Air Force

However, while the Pentagon, Project Blue Book, Nickell, and McGaha strove to diminish and dismiss the eyewitness testimony of the three "stable, reliable persons," they all failed to mention all of the other reliable witnesses that night—the dogs who don't howl and bark at temperature inversions, and the horses who don't raise "holy hell" and frantically kick their stalls because of conventional aircraft to which they had been accustomed. (Even decades later, the damage caused that night by the kicking horses was still plain to see.) And let us not forget all the crickets on that summer night in the surrounding woods mysteriously going silent.

Once again, these animal reactions support the human testimony that this was an event that was unusual, out of the ordinary, and frightening. Due to the horses, dogs, and crickets, the Incident at Exeter should be considered to have been real, and still very much unexplained.

***

If there is an incident in ufology that least needs an introduction, it must be the Betty and Barney Hill case. Their encounter the night of September 19, 1961, on the dark highways of New Hampshire with the craft and "beings" who "were somehow not human," has led to over 50 years of research, investigation, and controversy over what many consider to be the first extensively studied case of abduction.

For all its half century of media attention, however, few realize that there was a third witness in the Hill's 1957 Chevy Bel Air that night— Delsey, their dachshund.

Delsey had accompanied Betty and Barney on their trip to Niagara Falls and Montreal, and seemed healthy along the journey. However, after returning home to Portsmouth, New Hampshire, she began having respiratory problems and a fungal infection on her skin, for which she needed to be treated by a veterinarian.

Also, and perhaps even more telling, was that Delsey "had suddenly begun to whimper, shake, and move her legs, as if running in her sleep, following their September trip. Although most dogs exhibit this type of behavior from time to time, it was intense and persistent in Delsey."[3]

---

[3] Stanton Friedman and Kathleen Marden, *Captured*, (New Jersey: New Page Books, 2007) p.55

What prompted Delsey's severe nightmares upon returning from the trip? Had she simply been affected by all of the excitement of the journey, or had she been traumatized by the sight and presence of an unknown craft? Or, had something more happened to Delsey, as well as to Betty and Barney? Perhaps all three of them had been given very good reason to have nightmares.

Delsey, with Betty and Barney Hill

***

The following is another example where the dog involved barely even gets mentioned in the many articles written about this famous case.

Lara was a happy, 6-year-old Irish setter who loved to roam the woods of Dechmont in Scotland with her owner, forester Robert Taylor. On November 9, 1979, Taylor was walking up a familiar path in the woods, and Lara was joyously bounding around through the trees—when she suddenly came back to him visibly frightened. As the birds all fell silent, Lara began whining and growling at something only her dog senses could detect. Taylor was about to pick her up when he saw a bizarre object.

It was a "flying dome" made of "a dark metallic material with a rough texture like sandpaper" and was about 20 feet in diameter and hovering just above the ground. There was a strange, almost static electricity feeling in the air, and the acrid smell of "burning brakes." Then the really terrifying part began.

Two small spheres, resembling "sea mines" with their tubular protrusions, came from the craft and began rolling straight at him. These objects dug into his pants and began dragging him toward the larger craft. Then Taylor blacked out. Sometime later, he awoke on the ground to Lara "furiously barking." The craft were gone, but Taylor was dazed, scratched and bruised, and his clothes were torn. He staggered back to the truck, but it wouldn't start, and he and Lara had to walk a mile to get home.

Mrs. Taylor was certain her husband had been assaulted, due to his disheveled appearance and state of confusion, so she called the police and the doctor. They were stunned when Taylor finally explained what had attacked him, especially as he was known to be trustworthy, and not one to make up stories.

However, to their credit, the police did go to the scene and found physical marks in the ground that matched up with the story of something mechanical. That, coupled with his torn clothing and injuries, led the police to believe that something real had occurred; although what, was the million-dollar question. The case became known as the "only example of an alien sighting becoming the subject of a criminal investigation."

Naysayers had a field day with the case, deciding it was either a hoax, a stroke, an epileptic seizure, or a secret military operation. Skeptic Steuart Campbell had the most creative explanation—that Taylor had suffered a hallucination brought on by a "mirage of Venus."

What none of these skeptics address, however, are the reactions of the second witness, Lara, not to mention all the birds that fell silent. Lara had clearly been highly distressed and agitated before Taylor saw the craft. For five days after the sighting, both Taylor and Lara lost their appetites. Most importantly, Lara, the energetic, outdoor-loving dog, *refused to leave the house for months* after the incident! This was a dog who thoroughly enjoyed her daily adventures in the woods, but now she didn't even want to leave the house.

Would Lara have had these long-lasting reactions because her owner had a temporary hallucination due to a mirage of Venus? What had so traumatized this dog that she was too terrified to even leave the house for months? Skeptics can fabricate all of the wild excuses they want, but the tacit testimony of Lara speaks volumes to the fact that something extraordinary happened in the woods that day.

***

Another famous case that deserves at least a brief mention is the Falcon Lake Incident, which might not have even happened had it not been for some very agitated birds.

Stefan Michalak was a 51-year-old amateur geologist prospecting for quartz and silver at Falcon Lake, Manitoba, Canada, on May 20, 1967. Concentrating on his work and looking down, he was startled into looking upward by a sudden "explosion" of honking sounds from a large gaggle of geese who were obviously extremely disturbed by something in this normally tranquil setting.

That "something" happened to be two strange, metallic craft in the sky, one of which had landed. Thinking it must be an experimental vehicle from the United States, Michalak actually took the time to sit down and draw a detailed sketch of the object!

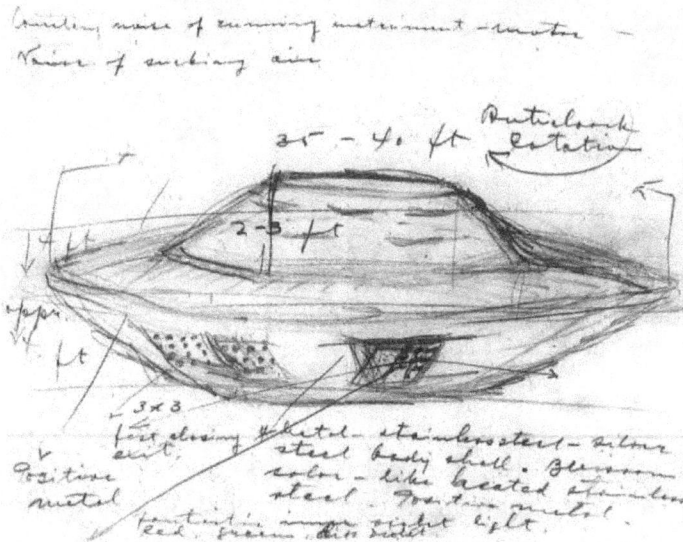

Michalak's sketch of the craft.

Still not alarmed, he approached the craft to see if the "Yankee boys" were having trouble and needed help, as he was a mechanic. A door was open, there was an intense light inside, and he heard voices. Calling out in English, Polish, German, and Russian, he received no response.

Unfortunately, the door then closed, the craft turned, and a blast of hot gas hit Michalak, setting his shirt and hat on fire, and burning a remarkable grid pattern into his chest and abdomen. The craft then lifted up and flew away, leaving him dazed, disoriented, and nauseated to the point of vomiting. He struggled to return home, and had to go to the hospital to have his burns treated. For several weeks after, Michalak suffered from a host of physical problems, including headaches, diarrhea, weight loss, and fainting spells. His body also emitted a strong smell of sulfur.

While the geese of Falcon Lake can't tell us exactly what frightened them so badly, they do provide a cautionary tale to human witnesses. Whatever it was that provoked their extreme sense of danger, perhaps we need to respect that sense and keep our distance from these craft.

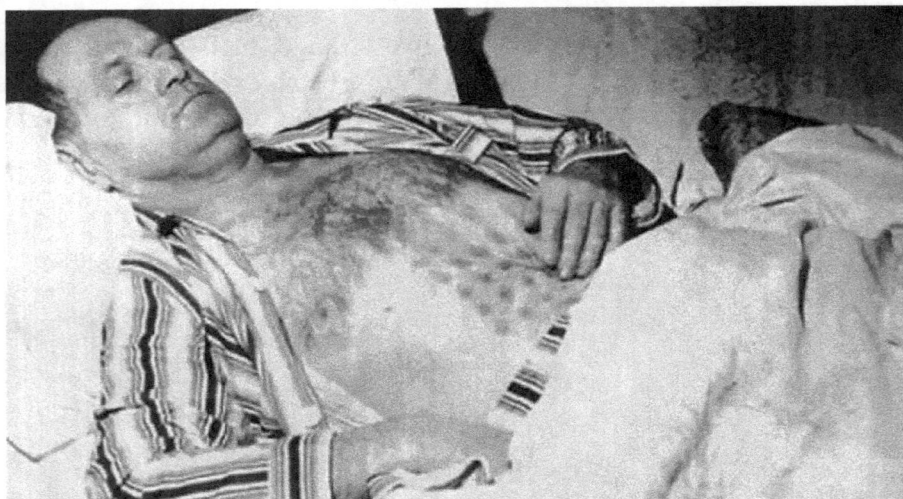

The grid pattern burns.

***

The following is not a case that is well known, but it is an excellent example of how an animal witness precludes the absurd official explanation of the sighting.

70

Every witness is valuable, but when that witness was a WWII photographic interpreter who served in the Middle East with the British Royal Air Force, and then went on to a career with British Overseas Airways Corporation (BOAC), one would think his testimony would be regarded as something very special. And when the craft in this sighting was in Dorset, England, "somewhere between the Winfrith Atomic Station and the Portland Underwater Defence Station"…"and about a mile inland from the USAF Communications Unit at Ringstead Bay,"[4] officials should have stood up and taken notice.

Angus Brooks had taken his 12-year-old German Shepherd and 4-year-old Dalmatian to the Moigne Downs on October 26, 1967. The dogs ran off to scamper about as usual, and at 11:25am, Brooks saw an unusual craft descend "at lightning speed" down to an altitude of no more than 200 or 300 feet. The craft was circular, and about 25 feet in diameter and 12 feet high. The really bizarre features were four, 75-foot-long, grooved "fuselages" that came out from the center and were able to rotate positions. These structures were about seven feet high and eight feet wide. Initially, one fuselage was in front, and three clustered in the back, but then they moved "to form four fuselages at equidistant position around [the] center chamber," in a shape reminiscent of a ceiling fan. Brooks said the entire craft appeared to be made "of a translucent material."

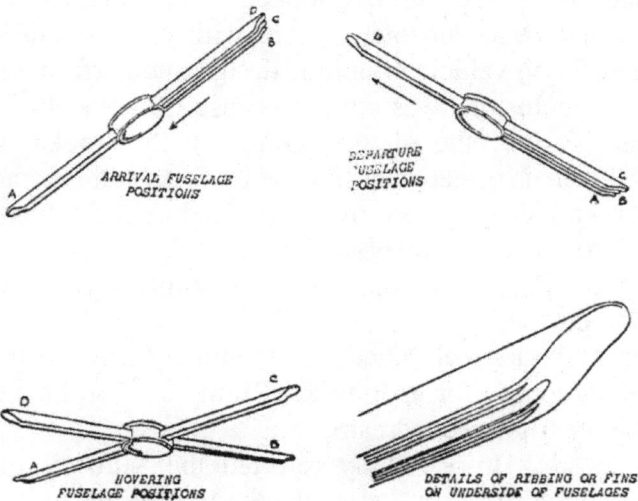

Sketch made by Angus Brooks.

---

[4] *Strange Effects from UFOs,* p.35-37

71

"For 22 minutes the strange object remained motionless in the sky," and "unaffected by very strong wind." However, Brooks' German Shepherd, who returned during the sighting, was not unaffected by the craft, and was, in fact, very "distraught."

"The dog was standing here and her ears were pricked straight up like she does when her ears heard sounds that she was worried about." The dog kept trying to coax him to leave, and would not respond to his commands of "sit" or "down," which was completely out of character for her. The craft flew off at 11:47am, and headed to the northeast, disappearing from view.

From that day, the four additional times Brooks returned to that spot with the dogs, his German Shepherd acted very "nervous," and her unusual behavior was noted by other witnesses. Unfortunately, we can't know how long this trauma-induced reaction would have continued, as the dog passed away just six weeks later of acute cystitis.

Brooks, with his extensive military and aviation background, knew he had witnessed something extraordinary, and dutifully reported it to the Ministry of Defence in London. Not surprisingly, the response was both absurd and disrespectful to a man of Brooks' credentials.

Investigator L.W. Akhurst wrote:

"We do not doubt that the experience which you have described was a very vivid one, nor have we overlooked your long association with aviation. However, we are unable to agree with your conclusion that you saw a controlled flying vehicle of unique design and performance."

Akhurst's conclusion was that because Brooks had a corneal transplant years earlier, the alleged craft, which Brooks was able to describe and sketch in great detail, was, in fact, nothing more than "a vitreous floater—a piece of loose matter (a dead cell) floating in the fluid of the eyeball." And it went downhill from there.

Probably realizing this was a flimsy and ridiculous excuse, Akhurst needed to add to his theory.

"However," Akhurst continued, "it is unlikely that the floater would have remained stationary for as long as 22 minutes," so he suggested the floater "could have triggered a dream state."

Needless to say, Brooks flatly rejected the Ministry of Defence's theory, and so should everyone else, thanks to that all-important second witness. Would a dog have been agitated, distraught, and on alert, because Brooks had a floater in the fluid of his eyeball? Would this German Shepherd have continued to act nervously every time she returned to the location of where Brooks had entered a floater-induced dream state? And what of the five other human witnesses who saw a similar craft in other parts of the country during that month of October? Perhaps the Ministry of Defence could turn a blind eye to this remarkable case, but both human and animal witnesses see through the nonsense.

***

Rendlesham.

That single name has spawned countless books and shows about the case that has come to be known as "Britain's Roswell." The events in late December of 1980 involving the RAF Bentwaters and RAF Woodbridge bases in Suffolk, England, both being used by the United States Air Force at the time, has arguably garnered almost as much attention as the original Roswell incident.

A comprehensive recap of the sightings on December 26 and 28 would fill this book, and because so much information about this case already exists, the focus here will be on the animal reaction aspect. Suffice it to say that military personnel on these two nights saw strange lights and a physical craft in Rendlesham Forest. Key figures in the sighting were Staff Sergeant Jim Penniston, who saw and touched the "craft of unknown origin," Airman John Burroughs, who was with Penniston, and deputy base commander Colonel Charles Halt, who recorded the famous audio tape, and wrote the memo that gained so much notoriety.

For all of the evidence, eyewitnesses, and detailed testimony, to this day, some skeptics claim that all of these highly trained military personnel were doing nothing more than running around the woods chasing the beam from the Orford Ness Lighthouse. This theory began very early on when local forestry worker Vince Thurkettle was filmed by the BBC in 1983. Witnesses like Col. Halt have understandably been frustrated over the years by the repeated insistence that it was only a beam from the lighthouse, but it seems no amount of testimony and evidence will sway the skeptics. However, where human witnesses fall short, could the animal witnesses prove that something truly out of the ordinary occurred those nights?

John Burroughs wrote that "The woods lit up and you could hear the farm animals making a lot of noises, and there was a lot of movement in the woods." Those noises were clearly ones of distress and alarm, as if the animals were desperate to get away from something.

Also, Jim Penniston noted that the "animals around us were in a frenzy" and there was "wildlife running by us."

On Colonel Halt's tape, he can be heard saying, "We're hearing very, very strange sounds out of the farmer's barnyard animals...They're very, very active, making an awful lot of noise." Some have suggested that what they were hearing was actually muntjac deer, which do make strange sounds. However, whatever animals were making the sounds, they all stopped at the same time, as Halt says, "The light's still there and all the barnyard animals have gotten quiet now." In his memo, Halt also states that on the first night two "USAF security police patrolmen" noted that "the animals on a nearby farm went into a frenzy."

Think about that wildlife and those farm animals for a moment. Surely, they were most likely all born in the area, and had lived their entire lives there. At the very least, the wildlife and farm animals must have spent many, many days—and more importantly, many nights—within range of the Orford Ness beacon. That beam incessantly sweeping through the woods and over the fields was something they saw every single night, all night long, day after day, month after month, and year after year.

Therefore, are we to believe that suddenly on these nights alone, every animal inexplicably became panicked by the lighthouse light, and decided, en masse, to make sounds of alarm and run for their lives?

No. That is an absurd idea. Something extremely unusual spooked these animals, and it cannot be a coincidence that it just so happened to occur when human witnesses were also experiencing something strange. No doubt the Rendlesham controversy will rage on long after all the witnesses are gone, but thanks to the numerous, unbiased animal witnesses, we can at least finally stick a pin in the Orford Ness lighthouse excuse.

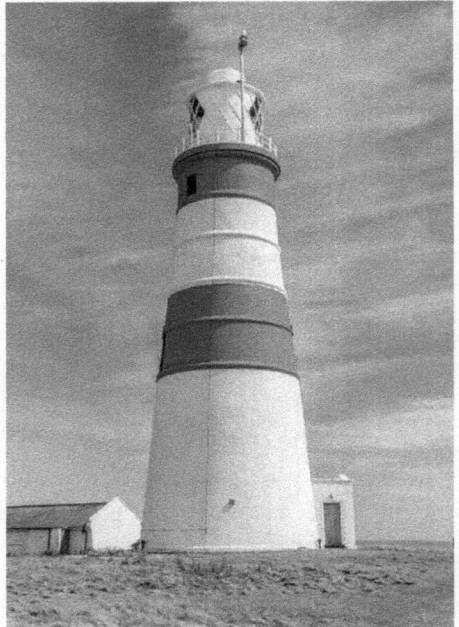

Part III
Casebook

There are hundreds of examples of animal reactions to UFOs, and the following is a sampling of some of the more interesting cases. They are arranged chronologically, and list the types of animals involved, location, and date.

Horses
Bonham, Texas, and Fort Scott and Fort Riley, Kansas
July and August, 1873

The Leavenworth Weekly Times, July 10, 1873
The people of Bonham, Texas, are in wonder over of the portentous sign of a large serpent which they have seen in the clouds. It signifies that if they do not swear off from the quality of the whisky they now are drinking they will die of snakes.

A more serious report stated:

The Fort Scott Kansas Monitor says a strange and remarkable phenomenon was observed at sunrise on Friday last. The sky was clear and the sun rose entirely unobscured. When the disk of the sun was about half way above the horizon the form of a huge serpent apparently perfect in form, was plainly seen encircling it, and was visible for some moments. The editor has the statement from two reliable witnesses, who are willing to make affidavit to the above.
The same serpent has been in Texas, as will be seen by the following from the Bonham Enterprise: a few days ago, a Mr. Hamlin, residing some 5 or 6 miles east of Bonham, saw something resembling an enormous serpent floating in a cloud that was passing over his farm. Several parties of men and boys, at work in the fields observed the same thing, and were seriously frightened. It seemed to be as large and as long as the telegraph poles, was of a yellow striped color, and seemed to float along without any effort. They could see it coil itself up, turn over, and thrust forward with its huge head as if striking at something, displaying the maneuvers of a genuine snake. The cloud and serpent moved in an

easterly direction and were seen by persons a few miles this side of Honey Grove. The question is what is it, and where did it come from.

Another report stated that the object in Texas repeatedly dived toward men in a cotton field, frightening both them, and the horses. "A team of horses ran away, and the driver was thrown under the wheels of the wagon and killed."

In Fort Riley, Kansas, at the "Same day an hour or so later, a similar object swooped down on Army troops on the cavalry parade ground, terrorized the horses to such an extent that the cavalry drill ended in a tumult."[1]

Singular Phenomenon.
(Fort Scott Monitor.)
A strange and remarkable phenomenon was observed at sunrise yesterday morning. The sky was clear, and the

—The people of Bonham, Texas, are in wonder over the portentious sign of a large serpent which they have seen in the clouds. It signifies that if they do not swear off from the quality of whisky they are now drinking they will die of snakes.—*Troy Chief.*

A Snake in the Sky.

The Fort Scott (Kan.) *Monitor* says a strange and remarkable phenomenon was observed at sunrise on Friday last. The sky was clear and the sun rose entirely unobscured. When the disc of the sun was about half way above the horizon the form of a huge serpent, apparently perfect in form, was plainly seen encircling

Seals
San Francisco, California
November 22, 1897

During the Mysterious Airship wave, a craft "flew over Cliff House and projected its powerful beam on Seal Rocks, causing the seals to dive frantically in the water."[2]

---

[1] Frank Edwards, *Stranger than Science*, 1959
[2] Lore and Denault, *Mysteries of the Skies*, p. 7

Ducks, Chickens, Mules
Conway, South Carolina
11:15pm January 29, 1953

After closing up his general store on Highway 701 for the night, 29-year-old Lloyd C. Booth headed for home about a mile away, where he lived with his parents. Relaxing with the newspaper, he suddenly heard one of the mules causing a commotion. At first, he wasn't too concerned, as "that mule was easily excitable."[3] However, when the other mule started making a racket, he took notice, as "that mule is hard to frighten and is always quiet."

Then "the chickens began squawking and a pen of ducks simply went wild with excitement."

He was especially concerned by all of this commotion in the barnyard as one of his cows had just died the night before under mysterious circumstances. The cow had shown no indication that it was sick, or that anything was wrong. And the Booth farm wasn't alone—in the past three weeks, 18 cows in the county had met the same unusual death, and two more would die that night on another farm just a mile away. Veterinarians examining the carcasses suggested they had all died of poisoning. Pigs were also dying in large numbers across the area. One farmer lost 75 pigs.

Concerned for the safety of his animals, Booth grabbed his Harrison and Richards .22 revolver and hurried outside to see if he could catch the culprit responsible for all of the apparent poisoning deaths. Bracing himself for a confrontation, he was completely unprepared for what he was about to encounter.

Only 10 feet above the pine trees was a craft that "was roughly 24 feet long. It was 12 to 14 feet across and 8 to 10 feet high. The front of it came up at about a 60° angle, the back at about a 40° angle, and it all blended in all around in a highly streamlined manner. Actually, it had the shape almost like a half an egg cut end to end...There were two glassed over areas in the front of the craft that might have been some kind of cockpit, although they were not like bubbles which our craft have."

It was a greyish color, and moving slowly enough that he was able to walk under it and get a very good look from all sides. He followed it for between 20 and 30 minutes, carefully studying the craft, which "had no visible means of support...no propeller...no exhaust." The only sound was a faint hum.

---

[3] *The Charlotte Observer Sun*, May 15, 1953, p. 8A

Having served in World War II and been trained in anti-aircraft, Booth was familiar with blimps, helicopters, and all forms of conventional aircraft. This was unlike anything he had ever seen before. One can only imagine the rising anxiety level of being alone in the woods with an unknown craft for such a long time. Finally, Booth felt there was only one thing he could do.

He raised his pistol and fired.

"I heard the bullet hit. It made a metallic noise and bounced off. I immediately fired again, but did not hear the bullet strike because almost instantaneously with the first shot, the object began making a much louder noise, like a big electric motor, and took off at a rapid rate of speed at an angle of about 65 degrees. It traveled much faster than any aircraft I have ever seen before and went completely out of sight."

'FLYING EGG' HUNTER AND PISTOL
Lloyd Booth in his store.
The Charlotte Observer,
March 15, 1953

Aware of the publicity and ridicule he would face, he only told his parents, as they had heard the gunshots. It was a week before he confided in a "lifelong friend and neighbor," who apparently lacked discretion, and the media circus began. Regardless of the fantastic nature of the craft, one thing the media could not question was Booth's honesty and character. From his pastor to the local newspaper editor, he was described as being a sober Methodist whose word was not to be doubted. And, he never sought publicity and never tried to profit from it in any way.

As Booth could not be discredited, officials sought a suitable excuse. Therefore, the Civil Aeronautics Administration decided it had been one of three Navy blimps that had been traveling from Georgia to Weeksville, North Carolina. They claimed Booth's description was an exact match for the gondola of a blimp. Even if the descriptions were similar to the gondola of a blimp, there was one small problem—the enormous blimp part was completely missing! Also, none of the Navy blimps reported being shot at, and no bullet marks were discovered on their gondolas. Not to mention, what would a Navy blimp be doing hanging around at treetop level over Lloyd Booth's farm for half an hour?

"I've seen many blimps and I've been in one," Booth said in response to the CAA's explanation. "I'd certainly know a blimp when I saw one 80 feet over my head...I watched the thing for 30 minutes."[4]

Then, of course, there were the reactions of the ducks, chickens, and mules. What had they sensed from this unknown craft that caused them to react with such fear? And while the mysterious deaths of over one hundred animals in the area cannot directly be linked to the unusual UFO, perhaps it was not a bad thing that Lloyd Booth summoned the courage to shoot at it and drive it off.

Cows, Horses, Dogs
Willow Grove, Victoria, Australia
7:10am February 16, 1963

The Auckland, New Zealand *Star* newspaper reported an incident that occurred the day before in Willow Grove, near Moe, Australia, 80 miles southeast of Melbourne. Farmer Charles Brew and his 20-year-old son, Trevor, were milking cows when they "saw an object descending through the rain to a height of between 75 and 100 feet" which agitated the cows and dogs.

It was about 25 feet wide and 10 feet high, with a six-foot antenna on top, in the shape of a battleship, like a "grey disc with a transparent band around its circumference and a number of scoop-like protrusions." The disc appeared to be metallic, with a pale blue underside, and made a "pulsating, whooshing sound as it revolved overhead."

"I thought it was going to land," Brew said, "but it suddenly shot off to the west at two to three times the speed of a jet and disappeared into a cloud."

How did the animals react to this craft? All of the animals on the farm reacted in extreme fear, and the *Star* reported that, "The cows turned somersaults and the horses reared in panic." Mr. Brew later denied the report about the somersaults, and clarified that he, in fact, said that his 150 cows did "everything bar somersaults. They certainly played up."

He did confirm that their reactions were violent, to the point where many of them broke out of the paddock. The fleeing cows had to be

---

[4] *Rocky Mount Telegram*, February 22, 1953, p. 5A

rounded up, but for several days they were so traumatized that they refused to go back into the paddock over which the UFO had briefly hovered.

Officials who interviewed Brew and his son found them to be very credible witnesses, and Brew even made a detailed sketch of the craft and included dimensions. However, despite the credible testimony, sketch, and animals' reactions, the official explanation was that it had been a tornado!

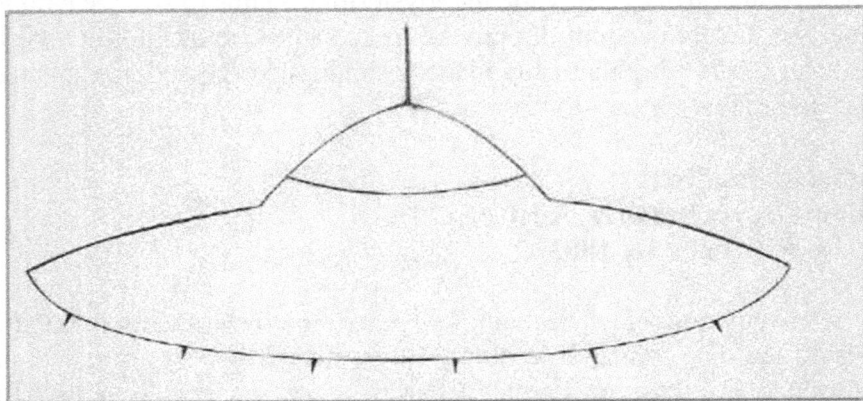

Sketch from the Royal Australian Air Force report.

And one more thing—*The Advocate* newspaper had an article about the sighting on February 20, along with something very unique. This is perhaps the only poem in existence mentioning animal reactions to UFOs!

From Mars to Moe
At Willow Grove, north west of Moe,
One starry summer's night,
A flying saucer 'peared on high,
And gave the cows a fright.
Don't scoff or scorn at Willow Grove,
Or throw jokes at its face,
For Willow Groves' not far from Moe
And Moe's the queerest place.

For we who've lived here long enough
Are not surprised one bit
That men from Mars should visit us,
And give the cows a fit.

The poem goes on—and on—from there, making fun of the politics and practices of the town, so this was the only relevant section. And as further proof of the author's sense of humor, he asked for anonymity as he is "strongly against capital punishment."

Dogs, Chickens
Trancas, Argentina
9:30pm on October 21, 1963

On the isolated Moreno farm, nine men, women and children witnessed "something like a small train, intensely illuminated,"[5] with figures moving around it by the nearby railway line. At first, they thought there had been some mishap or repair work going on, until half of the "train" shot away "skimming just above the ground." As this object sped away, they could "clearly" see that it was, in fact, "three circular objects." The other half of the "train" was also composed of three circular craft, each about 26 feet in diameter, with six porthole-like windows.

Two of the women bravely went into the garden to get a better view and one lit a lantern, and "as though in response to it, one of the brilliant beams of light emanating from one of the machines on or above the railway track at once turned from white to violet, and was switched round so as to play upon the two women in the garden. They immediately were overcome by suffocating heat, and tickling or tingling sensations in the body, and were obliged to run back into the house.

"The beam now turned and was playing steadily upon the house.

"Inside, the terrified family preceded to hide the children in different places in the house, and for the next 40 minutes they all cowered there in anxious silence. The temperature inside the house had risen greatly on account of the light beam, and the air was filled with the smell like sulfur, and everyone could feel the burning, needling, tickling sensations in their body.

"The family owned three very fierce dogs, but these, like the two dozen fowls outside, were powerfully affected by the beams of light. The family noticed that so soon as the beam shining into the rooms fell upon the dogs, the animals at once appeared to be listless and enervated. But then occasionally, when the beam fluctuated temporarily, or was playing

---

[5] *FSR*, 1966, V12 N1, p. 23-24

on another part of the house and grounds, the dogs seemed to come to life again and begin to growl."

At one point, "the old lady," 63-year-old Doña Teresa Moreno, "with great courage," went from window to window to observe what was happening, and saw that two craft were about 30-40 feet above the ground, and only 70 yards from the house. Later, five craft were near the house, illuminating the house and grounds with their "tubular" beams of light, some white, some "reddish-violet." There also appeared to be some "whitish gas" and "some sort of howling noise."

After a very long and frightening 45 minutes, the craft took off towards the Sierra de Medina mountains. Once gone, "The stupefied dogs now began to howl fearfully, and kept it up for some time."

Police later found that there were several other witnesses in the area, one of whom saw "six discs passing across the sky at between 10:15 and 10:20pm," just after they left the Moreno farm. A journalist who arrived the following day wrote that "the abnormal heat and sulphureous smell were still quite noticeable in the room."

Why were the "three fierce dogs" and the "two dozen fowls" seemingly sedated by the beams of light, but the human witnesses were not? What could it be that is capable of selectively incapacitating some animals, but not the human animals?

Cow
Hubbard, Oregon
7:30 am, May 19, 1964

"I went to put the cow out in the field," Mike Bizon told a local newspaper reporter. "Usually she can't wait to get out there...She was crosswise in the stanchion and seemed very nervous...she was bucking all the way."[6]

Bizon then saw a silver, four-foot-high square object with four legs in the middle of a wheat field.

"It started with a soft beep and started to go up," he told Marion County Deputy Sheriff Shirlie H. Davidson, who sent his official report to

---

[6] Donald E. Keyhoe, and Gordon I. R. Lore, Jr., 1969, *Strange Effects from UFOs*, Washington, D.C., NICAP, p. 53-54

NICAP. "It went up slow until it got to about the height of a telephone pole. Then it shot up just like a rocket..."

There was a distinct smell of gas fumes as the craft took off. In addition to the odor, a four-foot area of flattened wheat was found where the craft had been. The flattened area was seen by several people, including Deputy Davidson and an officer from Adair Air Force Base.

In Davidson's report, he wrote, "The grain appeared to have been pushed down by some object. Three particular spots were noted. These were spaced about 36 inches apart..."

Cows
Bridgwater, Somerset, England
Midnight October 1964

Fishing is not inherently dangerous, unless a herd of cattle is nearby and a UFO appears. The *Yorkshire Post* reported that four men were on a night fishing trip and shortly after midnight, "A mystery aerial object, thought to be a flying saucer" with red lights appeared, and:

"The men narrowly escaped being trampled to death by a herd of 50 cows terrified by the object. 'It was like all pandemonium let loose, and we hid behind a car in case the cows should sweep us into the water,'" one of the witnesses, Jim Sharman, said.

Sharman described the object as, "just like the red light on an aircraft, and as it got nearer it was so bright that it lit up the bank and surrounding fields. The light tapered to the rear...and when it got overhead it hovered, flashing on and off... It became so bright that the cows started making a heck of a noise and chased around the field."

The cattle were so panicked, that had the car not been there to protect the four men, they would have been seriously hurt, or worse.

Cows, Sheep
England, the Lake District
July-August, 1965

From the end of July through the first few weeks of August, there was a wave of sightings in the Lake District of England. Skeptics brushed off the numerous reports as everything from misidentified airplanes, to Venus, to "a plastic bag filled with gas," but the animal witnesses point to something highly unusual.

Michael Dean and his girlfriend, Molly Petherick, were enjoying their vacation at a hostel and were walking back to their room when they "heard a low droning noise, like an aircraft flying low and pulling at a very heavy load. Bullocks at a nearby farm started to kick up a din, which was unusual, as they had never made any noise before, despite the normal aircraft which passes over at regular intervals."[7]

The frightened couple saw a circular craft with a dome, which contained several illuminated, square "portholes." The craft revolved, and had flashing lights similar to a police car, only much larger.

"Miss Petherick's reaction was one of fear. This was something she had not experienced before, and she became shocked and stunned, and had to be taken back to the cottage in a state of collapse."

No one else there saw the object as they were all inside, but they "had heard the strange noise and the din of the bullocks."

During this wave, one farmer was drawn outside to investigate "the unusual noise" being made by his cattle. Another farmer described how a "flying cigar" caused his "cows and sheep to moan in a weird fashion." Yet another farmer said his "cattle and sheep had made a great deal of noise during the night," but he didn't bother to go outside to see what was happening.

With multiple eyewitnesses, over several weeks, across several counties, and numerous reports of highly agitated cattle and sheep making uncharacteristic sounds, surely, we can look at the Lake District wave as something more than a plastic bag filled with gas.

Dog
Tully, Queensland, Australia
January 19, 1966

On a typical Wednesday morning, George Pedley was driving his tractor when he heard a hissing noise. Initially thinking that the tractor tire was losing air, he then became alarmed when the hissing sound grew so intense, he knew it couldn't possibly be from his vehicle.

---

[7] *FSR*, Nov 1965, p.24-25

Nearby, over Albert Pennisi's farm, he saw a circular craft resembling two saucers joined together, only 120 feet away, and about 30 feet above a swamp. The craft dipped down, then took off at a remarkable speed and was gone.

He drove his tractor a little farther and could see a strange, circular depression in the tall swamp grass, approximately 30x20 feet in size. Pedley called the police, and Sergeant Moylan arrived. Moylan carefully inspected the area and said there were no other signs of tracks leading to the mysterious circle and had to conclude it was inexplicable, in terms of being made by either animals or conventional aircraft.

Albert Pennisi had even more to add to the account. He told Pedley that he believed his story about the flying saucer because "his dog suddenly went mad and bounded off towards the lagoon,"[8] where the craft was seen and the saucer "nest," as they came to be known, was found in the 5-foot-deep water. More Tully Saucer Nests were later discovered in the area.

One of the Tully saucer nests. Note the person to give an idea of the scale.

---

[8] *Sun-Herald*, January 23, 1966

Cattle
Yorktown, Iowa
2:10am April 23, 1966

The following is a case that's so unusual in its details that it may be difficult to believe, except for the fact of the reaction of the cattle and the trace evidence, which makes one realize something probably did happen on that foggy, rainy night when Ronald E. Johnson was awakened by a loud roaring sound. Looking to the south through his window, there was a 60-foot, cigar-shaped craft, only 50 feet from his house.

This craft is definitely one of the strangest on record, as Johnson claimed it had a "series of 17 to 20 long legs" upon which it landed. On the end facing the house, there was a deep red light that bathed everything in "a blood red glow."[9] There were two blue lights at the other end. Once on the ground, the roaring sound ended, but then there were a "series of loud, explosive, crackling noises at regular intervals, like gunshots, and the air was filled with an odor which the farmer recognized as resembling that of ozone."

Perhaps the most difficult part of this story to believe, is that after watching this incredible object for 20 minutes, Johnson simply went back to bed. He did get up again a short time later and found that the object was gone. When he awoke in the morning, he found that there were "distinct impressions in the ground where the craft had landed" that were circular, six inches in diameter, and spaced evenly in two rows, two and a half feet apart.

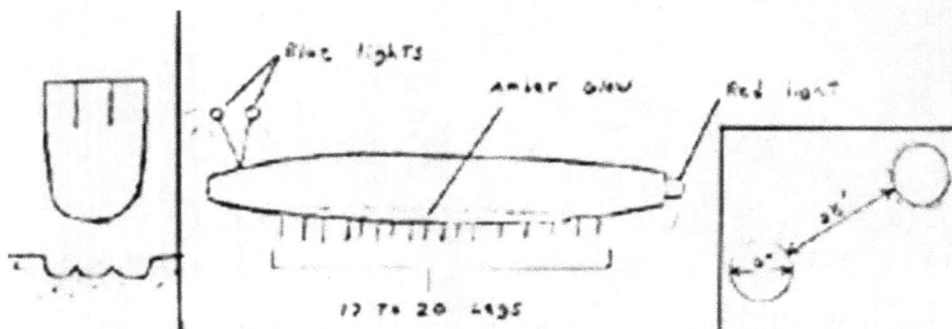

---

[9] NICAP Special Report, *UFOs: A New Look*, 1969, p.22.

"To the east of the landing impressions he found a second set of imprints, round on one side and square-edged on the other, and divided at the squared-off edge into three sections. These impressions were not made by his cattle, according to the farmer. One other peculiar detail was noticed: two power line poles appeared to have fresh depressions in the wood at regular intervals, as if it had been recently climbed. The poles were smeared with dirt. Wires rising on the poles had small, regular-spaced notches."

All of these details, along with the extreme cattle reactions, add credibility to this account, as Johnson found that the cattle had been "greatly disturbed during the night," and had all bolted to the far end of the pasture, "where, as he put it, they were still acting up considerably, and they refused to return to the farm that morning for feeding."

Deputy Sheriff Dick Hunt investigated, and saw the evidence for himself. Clearly, cattle would not act with such fear and refuse to eat if Johnson had a nightmare or hallucination. And neither would Johnson's imagination or misidentification of conventional aircraft have caused all the physical traces left behind.

Birds
Sawtry, England
5am April 28, 1966

Peter Rushton, a 9-year veteran with the Royal Air Force, was returning a van to London with Derek Robinson, and they had stopped on a side road to sleep. At about 5am, Rushton abruptly awoke because all the birds were screeching loudly. Rushton found that he was having trouble breathing, and there was an acidic taste in his mouth.

Opening a sliding shutter on the van window, he witnessed a large, round, orange craft in the sky nearby. Waking up Robinson, they both watched this craft hover, as the birds continued to screech in alarm. Then they saw a second, identical orange craft about a half a mile away. Suddenly, the two craft exchanged positions by moving at impossible speeds.

The craft then faded from sight, and the birds immediately fell silent. For at least a year after the sighting, Rushton continued to be disturbed by the experience, often awakening from nightmares in which he relived the incident.

Fish and Crows
Between Innis and Batchelor, Louisiana
4:30pm January 12, 9:30am January 13, 1967

There were a lot of "flying saucer" photos taken during the 1950s and 60s, and a gullible public, hungry for such things, couldn't get enough. Today, it doesn't take a sophisticated photo analyst to see that many of these photos were faked, and with the benefit of hindsight, we need to have an extra-critical eye on these cases, especially where not one, but three photos were taken over the course of two days. However, could the mention of animal reactions in this case add some credibility?

On the Raccourci Old River, a man—who wished to remain anonymous to avoid any publicity—was in his boat during a weekend camping trip. Glancing to the east, he saw a strange object in the "configuration of a shallow disc with domes on each side, one slightly smaller than the other,"[10] and managed to take this Polaroid photo of the disc-shaped craft before it disappeared.

That night, at about 9pm, he was checking his trotlines when he heard something that sounded "like a huge vacuum cleaner running full blast"[11] at a distance. In the complete darkness, he couldn't see anything, but given his sighting five hours earlier, he wondered if the sound and the craft were somehow related.

"The next morning, he was on the lookout for the object. The first thing he noticed was an unusually large number of dead fish along the river bank. He had noticed dead fish before but not so many."

Then around 9:30am, "he heard what sounded like a million crows 'chattering like crazy,' in a clump of trees across the river. He looked in that direction and there it was again, that same, or a similar object, in the sky above

---

[10] *UFO Investigator*, March-April, 1967, p. 6
[11] *APRO Bulletin*, March-April, 1967, p. 8

the opposite bank." Grabbing his camera, he managed to take two more photos before the craft took off straight up "like an elevator," and disappeared into the overcast.

According the local CBS network, who obtained the photos for analysis, the three Polaroids were deemed to be authentic by "two different agencies"[12] that examined them. Do all the dead fish and chattering crows further add legitimacy to this story, and therefore, these photos? If one were to make up a story, dead fish would certainly be a dramatic touch, but would the sound of alarm of what sounded "like a million crows" be a detail the average person would think to add?

Dogs
Milton, Indiana
6:30am February 22, 1967

Mrs. James Clevenger saw a strange sight out the kitchen window as she stood by the sink. For some reason, her collie had jumped up and threw himself against the window. Then he started racing around the backyard "barking and jumping"[13] in the most bizarre and puzzling manner. Her attention now drawn to the dog's inexplicable behavior; Mrs. Clevenger then saw something even stranger.

"It appeared as headlights of a car except there was only a solid light in an oval shape," she reported to NICAP. She then saw a row of white lights along the side of the craft. Clevenger let the poor, terrified collie in the house, where it promptly ran straight into the living room and hid. However, Clevenger was brave enough to try to get a better look, and ran barefoot out into the cold, dressed only in her pajamas. She was able to watch the object moving at a height of between 100 to 200 feet.

She ran back into the house and called a friend, Mrs. Judd Alford, who only lived about a quarter of a mile away. Alford's dog, a fox terrier, had also raced into the house and had hidden under a chair. Alford described the craft as "a circle of white lights some 200 or so feet in the air. The object appeared like a saucer to me."

About ten miles to the north in Hagerstown, Indiana, there were three more human witnesses to the craft that evening—Reverend and Mrs.

---

[12] *UFO Investigator,* March-April, 1967, p. 6
[13] *The Richmond Palladium-Item*, March 3, 1967

Leonard Lutz and their son, David. There was no mention if they had any pets that hid under furniture during the sighting.

Dog
Indiana
March 3, 1968

Throughout the eastern United States and Canada that night, many eyewitnesses saw bright objects racing across the sky. Wright Patterson Air Force Base, alone, received 70 reports—and those who actually report a sighting constitute a small fraction of the actual number of witnesses. The official excuse given by the North American Air Defense Command for this wave of sightings was that it was all just debris from the Soviet spacecraft *Zond 4*, which had launched the previous day.

However, while possibly some booster rockets reentered March 3, the *Zond 4*, itself, did not meet its fiery death in the Earth's atmosphere until March 8. Also, witnesses reported craft at treetop height, being close enough to see square windows, having a metal fuselage, or looking like a jet aircraft without wings.

Pilots were particularly good witnesses to this event, including reports from "an Eastern Airlines pilot over Connecticut, a United Airlines flight over Indiana, an American Airlines plane over Pennsylvania, a Piedmont Airlines pilot over Virginia, and an Air Canada pilot north of Toronto."[14] Rather than fiery debris at the fringes of space, most saw "what seemed to be a formation flight of several vehicles...only a few thousand feet above them."

Adding to the doubt cast on the *Zond 4* explanation, was the reaction of a dog in Indiana.

"A woman told how her dog, when the UFO passed over, lay between trash cans in her driveway and whimpered, 'like she was frightened to death.'"

Was it just the *Zond 4* debris, or was it something more that night? If it comes down to the statement of the North American Air Defense Command, versus the eyewitness accounts of numerous airline pilots, perhaps in this case, a single whimpering dog hiding between trash cans can help tip the scales.

---

[14] *Bangor Daily News*, July 5, 1968, p. 21

Rabbits
Hanbury, England
Between 5:30-5:45pm November 20, 1968

Milin and Doris Milakovic and their son, Slavic, "were driving through an area with fields on both sides of the road. They noticed first one rabbit and then several more scurry across the road from left to right. Looking in the direction of the fleeing rabbits, they saw a brilliant object in the field to the left."[15]

Stopping their car and getting out, for five minutes they watched as the object rose up and passed directly over them, producing the sensation of heat. "The object continued flying over the field to their right and hovered over a house about 100 yards away, where it quivered "like a jelly." It appeared to be "as wide as the house." They got a very clear view of the craft, and said, "In the lit dome area, several humanoid figures were seen moving, sometimes appearing to bend down as though looking at something below the rim."

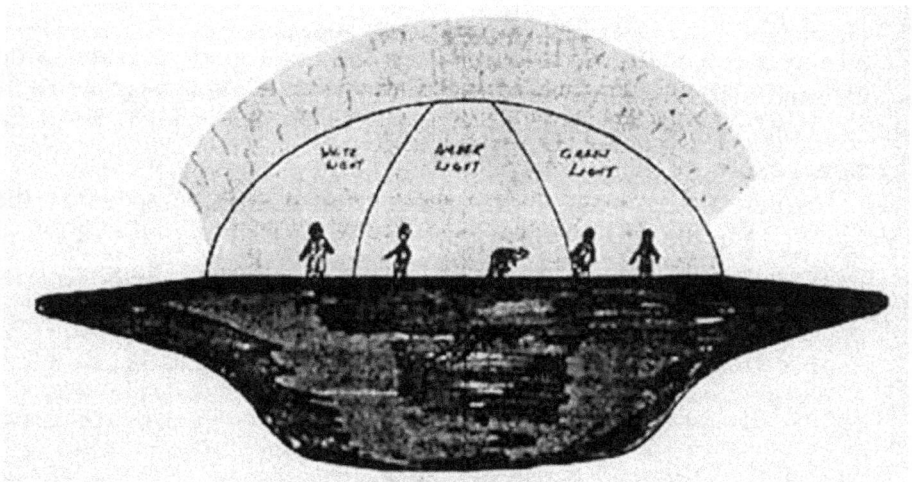

---

[15] Donald E. Keyhoe, and Gordon I. R. Lore, Jr., 1969, *Strange Effects from UFOs*, Washington, D.C., NICAP, pages 29-30

91

The light grew so intense that "Mr. Milakovic felt his eyes burning." Although normally "a brave man," Milakovic became so frightened that he "pushed his wife and son into the car and sped away."

Goats, Dogs
Saint-Martin-de-Londres, France
1am February 9, 1969

A 26-year-old goat farmer and restaurant owner, along with two other witnesses, saw a silent, red, glowing, 20-meter-long disc descend and hover over the field where he kept his goats. The owner flashed his car headlights at the craft, and the object went dark. When he turned off his lights, the craft began glowing red again. The sighting lasted several minutes before it disappeared at high speed.

The owner told investigators that "All the time the disc was there, my herd of goats were right under it or close by. They don't seem to have suffered any ill effects, but since that date, they have 'come on heat' at closer intervals, which is quite abnormal for goats. Moreover, I had an excellent male goat for breeding purposes, but I had to have him slaughtered, for he was no longer doing his job."[16]

Had the stress of the sighting affected the male goat's ability to do "his job," and also somehow altered the females' breeding cycles? (It is curious to note, that Ronnie Johnson stated that after the Delphos incident, the breeding cycles of their sheep also changed.)

The owner also noted other strange things happening after the sighting. For example, from that night on, his dogs all started to bark at 2am every morning. Also, he began hearing a strange tapping-type interference on his radio.

---

[16] Flying Saucer Review, January 1970, p. 1-2

Turkeys
Stover, Missouri
1:30am August 31, 1969

A husband and wife were asleep in bed when their flock of turkeys all called out in fright and alarm. Certain that a predator was after their flock, the couple jumped out of bed and rushed to the window. They were stunned to see an orange-red ball of light that was flat on the bottom, moving high in the sky in a northwesterly direction.

Compared to the moon, this object looked to be the size of a volleyball, and it was completely silent—although the witnesses did add that the turkeys were making such a racket that any sound could have been drowned out. A particularly strange feature of the craft is that it swung back and forth as it moved, almost like a pendulum. The entire sighting lasted about 3-4 minutes.

Dog
Rouge River, Medford, Oregon
1am Summer 1971

"After 48 years of silence," a woman finally decided to report her sighting to MUFON. The incident took place on a camping trip by the Hydro Electrical Station canal site with her husband and their "highly trained 2 years old pure black, fearless German Shepard dog, named Chandler." The dog was apparently so brave and tough he had "even survived a rattlesnake bite on his nose without a whimper."

They were doing some night fishing, and the dog was asleep next to the woman. Then "all of a sudden Chandler bolted to all four feet and started the strangest sound I ever heard from any dog then or to this day!...guttural-howl/growl-scream as if the dog might be in severe pain, attempted to crawl under my camp-chair where obviously he can not fit, then trying to cover his ears with his paws as if trying to block some ear piercing whatever?...after that, he was scraping his head on both sides of his ears along the sharp gravel in obvious horrific pain."

They then see a "huge pulsing-glow-fire-orange/red oval football-shaped craft above us with no apparent windows, seamless-looking solid material/construction and no visible means of propulsion...guessing 50

93

yards in length by 10 yards in height…hovering at the height of a 20-story building." The craft was spinning rapidly and Chandler seemed to be "acting worse for pain and screaming even more," even though the couple couldn't hear a sound.

They grabbed all of their gear and tossed it into the truck, and "strangely…Chandler jumped in the camper back with the gear as soon as the door was barely open, forcing his way inside using his head as a plow in a panic-fleeing-demeanor…I say strangely, as he never wanted to ride in the back of the camper, only wanted to sit in the truck cab with us, so I figured he must be as afraid as we are?!?"

Taking off so fast the truck tires were "spewing gravel," it took several minutes "before the dog recovered from his painful event and finally stopped panting."

Now divorced, the woman stated that immediately after the incident, her "husband blurts out in a scary, threatening tone, and commanded that 'I better not repeat this to anybody as he would deny it and call me crazy'…and actually 'with my dog as my witness,'" she affirmed that Chandler was a more "compelling evidence/witness of something strange in our midst than the opinion of my ex-husband."

A dog being more reliable than an ex-husband? Who can argue with that?

Dog
Conway, New Hampshire
7am October 1976

A man living on Passaconaway Road, 10 miles northwest of Conway, would jog in the beautiful White Mountain National Forest almost every morning with his white German Shepherd.

The dog was "running ahead of me as he often did," the witness reported to MUFON, "when I noticed he had stopped, was facing into the brush and was growling. His hackles were full up and his teeth bared."

The witness assumed the dog had spotted "a bear or big cat," and fearing an attack, tried to call off the dog, who "wouldn't budge…Then he seemed to totally cower down to the ground in total submission to whatever it was." The man "was petrified with fear," but then heard a "humming, buzzing sound and felt the rumble vibrating through my shoes," indicating it was some sort of a machine, and not an animal.

He then "somehow got my nerve up and very cautiously got up near my dog to see what he was staring at. I was shaking with fear but I felt like

I was being pulled in like a magnet would to a metal object. I felt like I had to go to it. As I got to my dog's side, I could hear him whimpering and see he was shaking. We were within 20 feet of this object."

The craft was "a grayish/black metal-looking object that had a military characteristic to it, but was only about 5 or 6 feet in height and seemed to have a crescent shape, not really a boomerang, but an elongated, rounded crescent shape with blue-greenish lights pulsating at regular intervals." It was about 30 feet in diameter and was resting on the ground.

"The ground vibrated and the humming sound was hypnotic and paralyzing. For a brief moment, maybe 2 seconds, the humming stopped, the lights stopped and my dog seemed to re-animate and slowly rose and backed off. I followed suit. We backed out about 4 or 5 steps and I took off running as fast as my body could push, my dog right in front—full speed."

A couple of days later, the man got up the courage to return—carrying his handgun this time. The dog "was most definitely still quite hesitant of approaching the site." Where the craft had been, he "could tell a large object of the shape I had seen had crushed the brush and ground cover down flat and looked swept over at the edges, like frayed leaves. After that I never jogged that same trail again. It was just too unnerving."

Dog
Michigan City, Indiana
10pm October 28, 1977

A couple was living on Georgia Avenue, which had access to the dunes on Lake Michigan. The husband took their black lab, Kahlua, for his nightly walk before bed. Leading the dog on the leash to the dunes, he removed the leash, but "she did not run into the dunes like normal," the witness reported to MUFON.

"She stood at my side with one front leg up, tail straight out, and the hair on her back stood up. She was pointing toward the beach opening of the dune trail we stood at. This was odd because I had never seen her do this stance since the day we brought her back from California, the year before."

The man saw a glow coming from the beach, and assumed his neighbors were having a bonfire party. Putting Kahlua back on the leash, they ran to the beach opening, expecting to see some friends and neighbors.

Access to the dunes on Georgia Avenue.

"The beach was deserted and no light was nearby. I looked left (West) toward the lake pier, and was amazed to see 3 spheres moving our way just above the gently rolling waves on the sand. I looked down at Kahlua and she was watching the approaching spheres to our left. The hair on my arms stood up and I noticed by feeling that Kahlua's back hairs were still up."

The three spheres moved slowly, at a walking pace, and the witness could "see the foam of the water being illuminated by the light the spheres gave off. It was similar to a fluorescent bulb light type of glow. The surfaces of the orbs were a yellow glow with small fingers of pink or orange lightning running around the entire globes. They reminded me of those science class balls that create static electricity. The glowing orbs were about 3 feet across and the same distance apart from each other as they traveled."

At the spheres' closest approach, they were only about 100 feet away, and slowly moved off out of sight. Upon returning home, the man's wife said they had been gone 15-20 minutes. The witness "called the police in Michigan City the next morning and they laughed and said it was probably airplane lights and hung up. I guarantee that this was no airplane lights or airplane."

Kahlua would undoubtedly agree.

Dogs, Sheep
Echuca Village, Victoria, Australia
3:30am March 7, 1978

Keith Basterfield's *Catalogue of the More Interesting Australian UAP Reports* includes the case of Mr. and Mrs. Gilham, who were awakened "by their dogs barking and disturbances among the sheep. The sheep had gathered around the house." Mrs. Gilham went outside and found everything lit up brightly.

"Then I saw this huge big light in the paddock. It was a burning, blazing, white light. It did not have a centre but it was like a great big cobweb. We watched it for about twenty minutes. It was hovering just a bit above the ground. It made my eyes ache...It was rather a nightmarish thing...a blazing, hovering, cobweb of white light. It wobbled, sort of shook."

In addition to the reactions of the dogs and sheep, there was trace evidence, "A four centimetres depression was left in the driveway where the light appeared. The dirt and gravel in the driveway were blown about." There is no mention of how long it took to calm the sheep and dogs, but Mrs. Gilham "suffered a severe headache for several days after the encounter."

Dog
Putnam Valley, New York
Fall 1984

"Joe," a local political figure, heard his golden retriever begin to whine one night. He assumed the dog wanted to go out, but when he opened the door to the backyard, the dog began to tremble and whine even more. Joe couldn't understand his dog's unusual behavior, and practically had to push the frightened dog out the door, where it immediately curled up into a tight ball and whined even louder.

Joe suddenly had a very "weird"[17] feeling and at that moment, the sky went dark. There, right above him, at an altitude of no more than 250 feet, was a solid black triangle at least the size of two football fields. Greenish-yellow lights lined the edges of the enormous craft, and the most disturbing part was that it was totally silent and completely motionless.

Joe yelled for his wife, who came running out and was stunned by a sight that was beyond anything she ever imagined. The massive triangle just sat above the house for at least five minutes, and they were both frightened by the incredible size of this unknown object so low in the sky. It finally began to slowly move away, but it took at least another 10 minutes to get out of sight.

Joe called the local police, who had been flooded with calls about the massive triangle. The next day the *Journal News* and Putnam County newspapers were filled with accounts of other witnesses seeing the same exact thing.

Cat
Wappingers Falls, New York
Between 8-9pm May 25, 1985

Lenore lived on Robinson Lane, and "went out to call in my cat" and noticed "it was just too dark for 8-9 pm in May."[18] Looking toward her next-door neighbor's house "there was a huge, huge circular dark thing just sitting there above the trees." Lenore went back in the house to call her neighbor, "not knowing she was in labor at the time with the twins." While "talking to her the thing started to slowly move more towards my yard and the cat ran in all pissed off." The craft just slowly moved out back over the property.

Horses
Bedford Hills, New York
May 1987

A woman who had a job taking care of expensive polo horses on an estate had an apartment above the stable, and her mother was staying with

[17] Linda Zimmermann, *In the Night Sky*, Eagle Press, 2013, p.196-97
[18] Linda Zimmermann, *In the Night Sky*, Eagle Press, 2013, p.313

her on a visit. For three consecutive nights, they "would wake and see light outside, as if the sunrise was very colorful, orange and red,"[19] then go back to sleep, and wake again to see that it "was still dark out and the sunrise would come later."

"This went on for 3 days, and after the third night, that next morning the Mt. Kisco newspaper, *The Patent Trader*, had an article on the front page about a ship that was seen just a country block from where I lived.

"Two people saw it up close and said it was the size of the Westchester Hospital, and it was hovering above the hay field at the corner of West Patent Rd. and Broad Brook Rd. I lived on Broad Brook.

"I could hear the horses, and for 3 nights in a row one or more of them had been kicking or thrashing in the stall so violently that it made me run down to the barn to see what was wrong. When I got down there, nothing was out of order. The horses were quiet, but alert, and just stood there looking at me.

"That Tuesday morning, I was getting a horse cleaned up for the owner, and he would not let me near his hind end. Since he was a polo horse and I tied his tail up all the time this was very unusual.

"Then someone noticed there was a little blood on his buttocks. Well, when I finally got him calmed down, I examined him to find a perfectly straight vertical laceration about 4" long and deep enough to cause a lot more blood than there was. Actually, there was only enough blood for us to notice that it was there, though most of the skin layers had been split open as if cut with a razor blade. A very sharp, clean, vertical slice.

"He had a laceration that required stitches, but he wasn't bleeding, so why would he need stitches if he wasn't even bleeding? I think the owner had opted out of calling the vet. We never found anything in the stall that would cut him that way, and he was afraid of people going behind him for a while after that. However, I remember that laceration healed up really fast.

"Boy, did it give me the creeps when I realized I ran down to the barn and someone or something could have been right outside the door; all three nights!"

---

[19] Linda Zimmermann, *More Hudson Valley UFOs*, Eagle Press, 2017, p. 176-77

Dogs
Congers, New York
Summer 1988

Kirk and three friends were enjoying a clear summer evening at Lake DeForest, when they saw what appeared "to be a small cluster of red and green lights, sort of 'curled' into a tight spiral, like a snail."[20] As the friends continued watching "the spiral begins to very slowly unravel into a dancing or undulating chain of lights, now moving north. The only sound was that of dogs howling crazily far away in the distance."

The craft approached so closely, they thought it was going to land, but it began to move away, while it seemed "like all the dogs of Rockland (County) are still going out of their skulls."

Even more than 30 years later, Kirk says, "I remember that night like it was yesterday, and it is among the most important, terrifying, bizarre, and beautiful experiences of my life."

Kirk's depiction of the craft—reminiscent of the 1873 Kansas and Texas "serpent" sightings?

---

[20] Linda Zimmermann, *More Hudson Valley UFOs*, Eagle Press, 2017, p. 227-37

Birds and Squirrels
Glen Allan, Mississippi
7:30am January 18, 1989

A man who had been a pilot with the Navy in WWII was in a tree stand hunting deer, when he saw a silver, cylindrical object. The sky was clear and the object was very reflective. He described the craft as being the size of a Boeing 747, and moving about 300 m.p.h. at an altitude of 1500 feet. The craft was completely silent. It passed directly overhead, and the entire sighting lasted about one minute.

While the cylinder was in sight, all of the birds and squirrels froze in place and were silent. Once the craft had gone, the animals began moving again and making their normal sounds.

Dogs, Insects, Birds
Cornwall, New York
Summer 1992

It was coming to the end of one of those glorious summer weekend days in the Hudson Valley of New York, when Kathy's kids came running inside yelling about a UFO. At first, she thought it was a joke, but quickly realized they were genuinely upset and excited.

Rushing outside, there was an enormous circular craft so close to the roof that she "thought it was going to crush the house." Many dozens of people gathered from the neighborhood to stand and watch in amazement at this disc-shaped craft with its bright, multi-colored lights, silently hovering so very, very close, before ever so slowly moving off.

However, this is not the most remarkable part of this case, as something occurred which sets it apart as one of the most baffling cases in ufology. For while the majority of witnesses would swear in a court of law that this was a circular object with colorful lights, there were people standing shoulder to shoulder with them who would swear that the craft was a solid black triangle with a single white light in each corner!

How is this even possible? How could people be standing in the same place, looking at the same object, yet seeing something so completely different? Who was right, or was everyone wrong and it was something completely different? Almost 30 years later, the various witnesses still steadfastly stick to their original assertions of it either being a disc or a triangle. There are more witnesses, however, that speak very strongly against this being some sort of hallucination or mass hysteria.

None of the dogs in the neighborhood were "barking and all the insects and birds were quiet for days."

Kathy's dog, Hooch, was a one and a half year-old, big, very strong, brave, and protective German Shepherd/Doberman/Husky/Rottweiler mix, yet, during the sighting, he hid under the bed the entire time. He was subsequently sick for 3 days and would not eat or drink—only taking water on the fourth day and then finally food six days later. For a couple of months afterward, "He did not like to be outside at night. And he would stay close to the kids." Hooch also appeared uneasy inside the house, as though "he would feel something in it."

What can appear as both a disc and a triangle, silence insects and birds for days, and traumatize a brave dog for months? This is the quintessential case of a UFO, as well as, animal reactions.

Kathy's sketch of the craft just above the roof. Note the military helicopters and jet that arrived after the object had gone.

Dogs
Tuttle, Oklahoma
9pm January 21, 1993

In this MUFON case, a mechanic and his brother witnessed a triangular craft with very bright white blinking lights in each corner. The tip of the triangle was pointed upwards, and the two men estimated it was at an altitude of about 500 feet and a mile away. While the hair on the arms of the mechanic stood on end, "The dogs were going crazy in the whole town, barking and causing quite a disturbance."

Dogs, Crickets, Toads
Wolcott, New York
Midnight May 1, 1994

In this MUFON case, a man went out on his deck to have a cigarette, and looked toward the direction his dog was facing as it was barking. Then "Some gigantic craft appeared out of nowhere just above this barn with such bright lights of the likes that I have never seen before. And it was as if this thing took immediate control over me and my dog quit barking and all the crickets and toads quit chirping, everything was silent and I couldn't move anything except my eyes."

The craft was also silent, and by the time it vanished, his cigarette had burned all the way down, which the witness estimates was a period of about eight minutes.

Dogs
White Lake, Michigan
8:30pm July 15, 2009

A 14-year-old boy was taking a walk and noticed "every single dog in the neighborhood was barking nonstop. That is what got my attention, I'm thinking... what's up with that, they never act like that." He then saw "an orange glowing ball of light" which "sat still for 30 seconds" before it slowly began to descend, at which point "the dogs are going even more nuts."

The boy and three friends watched in amazement as the orange light went down below the tree line. They assumed it had landed, but when they went to the site, they were unable to find any sign of it.

Dogs
Pine Bush, New York
9pm January 4, 2013

Carlos Torres' dogs were outside barking in a manner which made him go outside to see what was upsetting them and try to quiet them down.

"There was something huge in the sky that blocked out all the stars. It was a solid rectangle bigger than the house, and it was moving very slowly and silently. If you put your hands up you couldn't have covered the whole thing. You couldn't have covered it with three hands."[21]

The craft was at an altitude of only 200-250 feet and was silent, and moved so slowly it took two minutes to travel out of sight. Carlos and his wife were naturally "shocked" by what they were witnessing, and their 6-year-old daughter was terrified and was "freaking out." A short time earlier a few miles away, two other adult witnesses saw the same rectangular craft as it was headed toward Pine Bush.

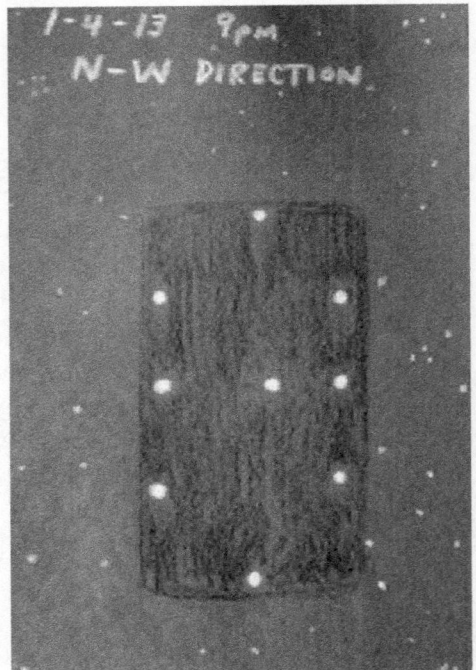

Carlos Torres' sketch of the craft.

---

[21] Linda Zimmermann, *Hudson Valley UFOs*, Eagle Press, 2014, p.53

Dogs, Cows
DeRidder, Louisiana
11pm Thursday, April 5, 2018

MUFON Case: "My dogs began growling in the living room. I live on a farm and the cows in the pasture across the street were mooing loudly. I went to the large picture window in the living room facing west. I looked for coyotes or other animals. My dogs were looking out the glass of the door and the window. The cows were looking up. I observed a huge round white light traveling slowly towards my direction."

For 15 minutes, the light grew in intensity and pulsated. When it suddenly disappeared, "the cows quieted, and my dogs stopped growling."

Dogs, Cats
Princeton, West Virginia
1:30am June, 22, 2018

In this NUFORC case, a woman's dogs and cats acted out of character during a triangle sighting. One dog was scratching at her bedroom door to get in, and her "German Shepard mix was acting super nervous and panting really hard." She went to use the bathroom, "and both dogs and one of the cats followed me, the dog that was digging down the door and even tried to jump on my lap."

At this point, she leashed the dogs and took them outside, where she saw a silent craft moving south. Feeling uneasy, she watched it move away and returned inside.

"The dogs calmed down for the most part but were still acting on guard. I have no explanation for the way they were acting. Both dogs and one of the cats acted weird. The other inside cat I couldn't find, our outside cat I couldn't find, and my ferret was in the bedroom with me asleep."

Interesting to note, that while the dogs and cats acted strangely, the ferret slept through the incident. Deafness is a common trait in ferrets.

Conclusion

The reactions of animals are not inconsequential, nor merely trivial curiosities. Animals as varied as insects, fish, birds, and mammals exhibit both short and long-term effects from UFO encounters. Such reactions have been catalogued around the world for generations.

Immediate reactions are overwhelmingly those of fear, often to the point of complete panic. Even after—sometimes long after—the event is over, however, those reactions can linger to the point of altering an animal's behavior for days, weeks, or even months. For example:

> ➤ Canaries wouldn't sing for 2 days
> ➤ Cows wouldn't give milk for a day or a week, and/or gave a significantly reduced amount of milk for many days
> ➤ Dogs refused to go outside for days to months, hid under beds and furniture, displayed extreme nervousness or distress when returning to a site
> ➤ Sheep kept jumping out of their pen for a week, cattle refused to return to a corral or field
> ➤ Insects and birds remain silent for days
> ➤ Animals refuse to go to the site of a landing for days, or forever

These lingering reactions suggest significant trauma brought on by something real and out of the ordinary. Astronomical objects, conventional aircraft they experience on a daily basis, and humans' flights of fancy do not account for their reactions.

Therefore, we need to look much more closely at animal reactions and:

A. Reevaluate existing cases based upon the animal reaction evidence.
B. Expand current investigations to fully explore all animal reaction aspects.
C. Create instrumentation that mimics or exceeds animals' senses for future investigations.

106

Wonderful new technology is now being deployed in the field of ufology, so in some aspects the future is looking very bright for research. As field investigators begin to acquire these high-tech instruments, they may also want to include something else—perhaps a four-legged sensing device that has constantly proven its remarkable sensitivity.

Perhaps man's best friend, is also one of his best research assets.

Index

About the Author

Linda Zimmermann is a former research scientist and an award-winning author of over 30 books on science, history, and the paranormal, as well as several works of fiction. Linda has been honored with multiple Best Author awards, and the Best Radio Personality award in the Hudson Valley of New York. She is a member of the Scientific Coalition for UAP Studies.

She enjoys lecturing on a wide variety of topics, and has spoken at the Smithsonian Institution, Gettysburg, West Point, the International UFO Congress, the MUFON Symposium, the Northeast Astronomy Forum, and national Mensa conventions. Linda has also made numerous appearances on radio and television.

Linda is the host, creator, and writer of three podcasts: *UFO Headquarters*, *Murder in the Hudson Valley*, and her personal favorite, *Science Friction Theater,* which is a spoof of 1950s' sci-fi movies.

She is a lifelong NY Mets and NY Giants fan, so don't even think of trying to call her when a game is on.

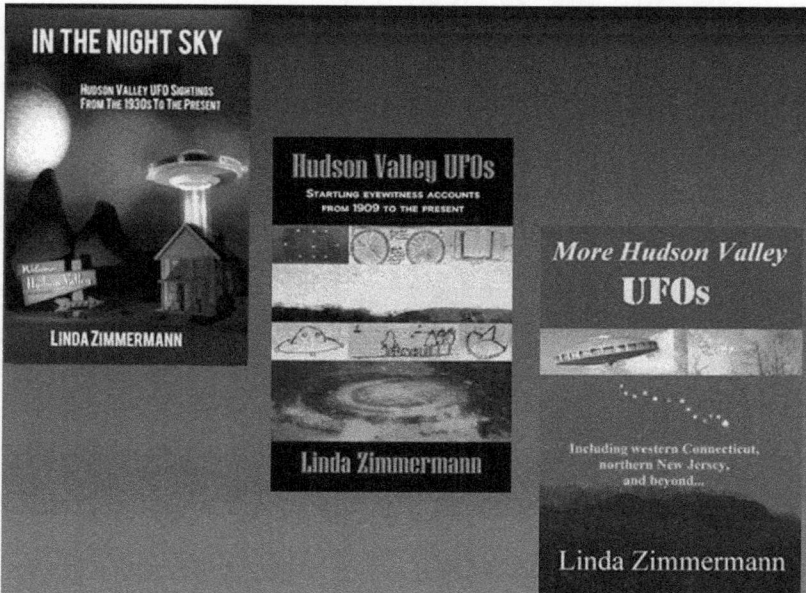